T0289602

Design of Integrated Circuits

Design of Integrated Circuits

Joseph Taylor

NY RESEARCH PRESS

P R E S S

New York

Published by NY Research Press
118-35 Queens Blvd., Suite 400,
Forest Hills, NY 11375, USA
www.nyresearchpress.com

Design of Integrated Circuits
Joseph Taylor

International Standard Book Number: 978-1-64725-429-2 (Hardback)

Cataloging-in-Publication Data

Design of integrated circuits / Joseph Taylor.
p. cm.
Includes bibliographical references and index.
ISBN 978-1-64725-429-2
1. Integrated circuits. 2. Integrated circuits--Design and construction. I. Taylor, Joseph.
TK7874 .D47 2023
621.381 5--dc23

Contents

Preface

Every book is a source of knowledge and this one is no exception. The idea that led to the conceptualization of this book was the fact that the world is advancing rapidly; which makes it crucial to document the progress in every field. I am aware that a lot of data is already available, yet, there is a lot more to learn. Hence, I accepted the responsibility of editing this book and contributing my knowledge to the community.

Integrated circuit (IC) design refers to a branch of electronics engineering that deals with the specific logic and circuit design methods required for creating integrated circuits. They are made up of miniaturized electronic components that are assembled into an electrical network on a monolithic semiconductor substrate through photolithography. IC design has been primarily categorized into two types including analog and digital IC design. The digital IC design is generally used for the production of components like FPGAs, digital ASICs, microprocessors and memories. Digital design emphasizes on the maximization of circuit density, logical correctness, and strategically placing circuits to route timing signals and clock efficiently. Analog IC design includes the specialties of RF IC design and power IC design. Analog IC design is used for fabricating oscillators, op-amps, phase locked loops, active filters and linear regulators. The topics included in this book on the design of digital circuits are of utmost significance and bound to provide incredible insights to readers. It presents this complex topic in the most comprehensible and easy to understand language. This book is a resource guide for experts as well as students.

While editing this book, I had multiple visions for it. Then I finally narrowed down to make every chapter a sole standing text explaining a particular topic, so that they can be used independently. However, the umbrella subject sinews them into a common theme. This makes the book a unique platform of knowledge.

I would like to give the major credit of this book to the experts from every corner of the world, who took the time to share their expertise with us. Also, I owe the completion of this book to the never-ending support of my family, who supported me throughout the project.

Joseph Taylor

CMOS: Inverters, Logic Gates and Timing Metrics

1.1 Introduction to CMOS Inverter

CMOS Inverter

CMOS inverters and adaptable MOSFET inverters are used in chip design. They operate with very little power loss and at relatively high speed. The CMOS inverter has logic buffer such that its noise margins is large in both the low and high states.

A CMOS inverter contains PMOS and NMOS transistors connected at the drain and gate terminals, a supply voltage V_{DD} at the PMOS source terminal and ground connected at NMOS source terminal, where V_{IN} is connected to gate terminals and V_{out} is connected to drain terminals.

CMOS does not contain resistors, which makes more power efficient than a regular resistor MOSFET inverter. As the voltage at input of CMOS device varies between 0 and V_{DD}, the state of NMOS and PMOS varies accordingly if each transistor is activated by V_{IN}.

MOSFET	Condition MOSFET	State of MOSFET
NMOS	$V_{gs} < V_{tn}$	OFF
NMOS	$V_{gs} > V_{tn}$	ON
PMOS	$V_{sg} < V_{tp}$	OFF
PMOS	$V_{sg} > V_{tp}$	ON

The above table explains when each transistor is turned on and off.

When V_{IN} is low, the NMOS is OFF, while the PMOS stays ON then it instantly charge V_{out} to logic high.

When V_{IN} is high, the NMOS is on and the PMOS is "off" taking the voltage at V_{out} to logic low.

DC Transfer Characteristics

The operation of the CMOS inverter can be divided into four regions. They are:

- Region A, the NMOS transistor is OFF and the PMOS transistor pulls the output to V_{DD}.

- Region B, the NMOS transistor starts to turn ON, pulling the output down.

- Region C, both transistors are in saturation. Those ideal transistors are only in region C for $V_{in} = V_{DD}/2$ corresponding to infinite gain. Transistors have finite output resistances on account of the channel length modulation and finite slopes over a broader region C.

- Region D, the PMOS transistor is partially ON.

- Region E, it is completely OFF, leaving the NMOS transistor to pull the output down to GND.

- The inverter's current consumption is zero, when the input is within a threshold voltage of the V_{DD} or GND rails. This feature is important for low power operation.

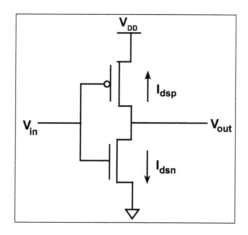

Region	Condition	p-device	n-device	Output
A	$0 \leq V_{in} < V_t$	Linear	Cutoff	$V_{out} = V_{DD}$
B	$V_t \leq V_{in} < V_{DD}/2$	Linear	Saturated	$V_{out} > V_{DD}/2$
C	$V_{in} = V_{DD}/2$	Saturated	Saturated	V_{out} drops sharply
D	$V_{DD}/2 < V_{in} \leq V_{DD} -\| V_{tp}\|$	Saturated	Linear	$V_{out} < V_{DD}/2$
E	$V_{in} > V_{DD} -\| V_{tp}\|$	Cutoff	Linear	$V_{out} = 0$

The equations are given both in terms of V_{gs}/V_{ds} and V_{in}/V_{out}. As the source of the NMOS transistor is grounded, $V_{gsn} = V_{in}$, $V_{dsn} = V_{out}$. As the source of the PMOS transistor is tied to V_{DD}, $V_{gsp} = V_{in} - V_{DD}$ and $V_{dsp} = V_{out} - V_{DD}$.

Relationship Between three Regions of CMOS Inverter

	Cutoff	Linear	Saturated
NMOS	$V_{gsn} < V_{tn}$	$V_{gsn} > V_{tn}$	$V_{gsn} > V_{tn}$
	$V_{in} < V_{tn}$	$V_{in} > V_{tn}$	$V_{in} > V_{tn}$
		$V_{dsn} < V_{gsn} - V_{tn}$	$V_{dsn} > V_{gsn} - V_{tn}$
		$V_{out} < V_{in} - V_{tn}$	$V_{out} > V_{in} - V_{tn}$
PMOS	$V_{gsp} > V_{tp}$	$V_{gsp} < V_{tP}$	$V_{gsp} < V_{tp}$
	$V_{in} > V_{tp} + V_{DD}$	$V_{in} < V_{tp} + V_{DD}$	$V_{in} < V_{tp} + V_{DD}$
		$V_{dsp} > V_{gsp} - V_{tp}$	$V_{dsp} < V_{gsp} - V_{tp}$
		$V_{out} > V_{in} - V_{tp}$	$V_{out} < V_{in} - V_{tp}$

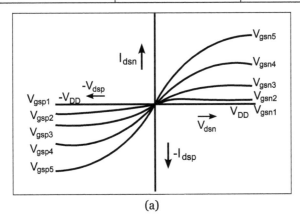

(a)

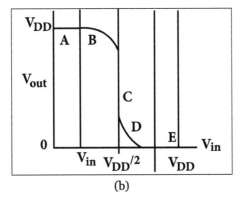

(b)

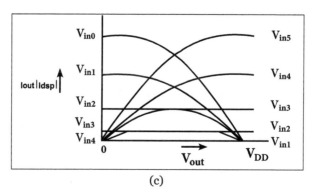

(c)

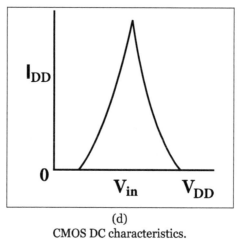

(d)

CMOS DC characteristics.

The plot shown above contain I_{dsn} and I_{dsp} in terms of V_{dsn} and V_{dsp} for various values of V_{gsn} and V_{gsp}. The possible operating points of the inverter, marked with dots, are the values of V_{out}, where $I_{dsn} = |I_{dsp}|$ for a given value of V_{in}.

These operating points are plotted on I_{out} vs. V_{in} axis. The supply current I_{DD} also plotted against V_{in} showing that both transistors are momentarily ON as V_{in} passes through voltages between GND and V_{DD}, resulting in a pulse of current drawn from the power supply.

Temperature Dependence of VTC of CMOS Inverter

Temperature also has an effect on transfer characteristics of an inverter. As the temperature of an MOS device is increased, the effective carrier mobility, μ decreases.

This results in decrease in K_n value, which is related to temperature T both VT_n and VT_p, decrease slightly as temperature increases and the extent of region A is reduced while the extent of region E is increased. Thus, the overall transfer characteristics shift to the left as temperature increases.

It is found that if the temperature rises by 50°C, the thresholds drop by 200 mV each. This would cause a 0.2 V shift in the input threshold of the inverter.

$$K_n \propto T^{-1.5}$$

$$I_{DS} \propto T^{-1.5}$$

Supply Voltage Scaling in CMOS Inverters

The overall power dissipation of any digital circuit is a strong function of supply voltage V_{DD}. The reduction of the power supply voltage emerges as one of the most widely practiced measures for low power design.

The second order effect, such as sub threshold conduction, in which CMOS inverter will continue to operate correctly with a supply voltage which is as low as the following limited value.

$$V_{DD}^{min} = VT_n + |V_{TP}|$$

Such that the correct inverter operation will sustain one of the transistors conduction in input voltage. The VTC of a CMOS inverter obtained with different supply voltage levels are given below.

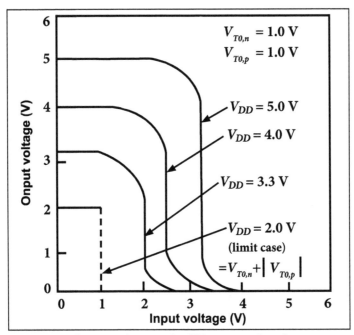

VTC of a CMOS inverter for different supply voltages.

If the supply voltage of CMOS inverter is reduced below the sum of two threshold voltages, VTC will contain a region in which none of the transistors is conducting. The output voltage level within the region is determined by the previous state of the output since the previous output level is always preserved at the output node. The VTC exhibits a hysteresis behaviour for very low supply voltage levels.

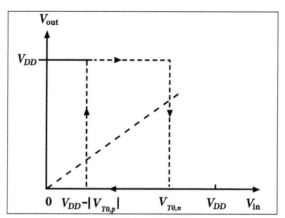

Hysteresis behaviour exhibited by VTC of CMOS inverter.

1.1.1 Introduction to Static CMOS Design

Conventional static CMOS circuit design techniques and designs are based on tri-state gates and pass transistors. These designs are static, since they do not require a clock signal for combinational circuits. Hence, if the circuit inputs are stopped, then the circuits retain their output state as long as power is maintained.

Advantages:

Dynamic CMOS logic families do not have this property.

- Fewer transistors result in smaller input capacitance, presenting a smaller load to previous gates and therefore faster switching speed.

- They use fewer transistors and therefore less area.

- Gates are designed and transistors are sized for fast switching characteristics. High performance circuits use these families. The logic transition voltages are smaller than static circuits requiring less time to switch between logic levels.

Disadvantages:

The disadvantages of dynamic CMOS circuits.

- Clock circuitry runs continuously, drawing significant power.

- Each gate needs a clock signal that must be routed through the whole circuit. This requires precise timing control.

- Dynamic circuits are more sensitive to noise.

- The circuit loses its state if the clock stops.

- Clock and data must be carefully synchronized to avoid erroneous states.

1.1.2 The Dynamic Behavior Delay

CMOS Steady State and Dynamic Electrical Behavior

- Resistance & Capacitance Estimation
- DC Response
- Logic Level and Noise Margins
- Transient Response
- Delay Estimation
- Transistor Sizing
- Power Analysis
- Scaling Theory

Resistance Estimation

$$R = (\rho / t)(L / W)$$

where,

ρ - Resistivity

t - Thickness

L - Conductor length

W - Conductor width

Sheet resistance is given by,

$$R_s = \Omega / sq. \text{ (or) } R_s = \rho / t$$

Thus,

$$R = R_s (L / W)$$

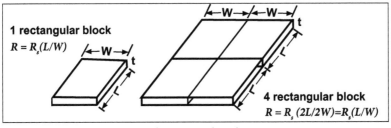

Resistance Estimation.

Capacitance Estimation

The switching speeds of MOS circuits are heavily affected by the parasitic capacitances associated with the MOS device and interconnection capacitances. The total load capacitance on the output of a CMOS gate is the sum of following:

- Gate capacitance

- Diffusion capacitance

- Routing capacitance

Understanding the source of parasitic loads and their variations is essential in the design process.

Diffusion Capacitance

Diffusion capacitance C_d is proportional to the diffusion to substrate junction area.

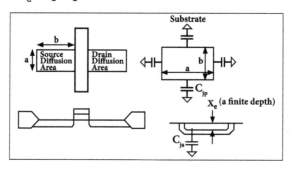

Diffusion Capacitance,

$$C_d = C_{ja} \times (ab) + C_{jp} \times (2a + 2b)$$

Where,

C_{ja} - Junction capacitance per micron square

C_{jp} - Periphery capacitance per micron

Junction Capacitance

Semiconductor physics reveals a PN junction automatically exhibits capacitance due to the opposite polarity charges involved. This is called junction or depletion capacitance and is found at every drain or source region of a MOS.

The junction capacitance varies with the junction voltage. It can be estimated as,

$$C_j = C_{jo} \left(1 - \frac{V_j}{V_b} \right)^{-m}$$

Where,

C_j - Junction voltage (negative for reverse bias)

C_{jo} - Zero bias junction capacitance $(v_j = 0)$

V_b - Built-in junction voltage ~ 0.6

Single Wire Capacitance

Routing capacitance between metal and substrate can be approximated using a parallel plate model.

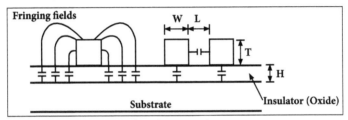

Single Wire Capacitance.

In addition, a conductor can exhibit capacitance to an adjacent conductor on the same layer.

Multiple Conductor Capacitances

Modern CMOS processes have multiple routing layers. The capacitance interactions between layers can become quite complex. Multilevel layer capacitance can be modeled as below:

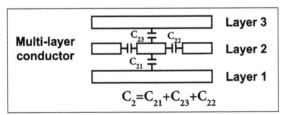

Multiple Conductor Capacitances.

Noise Margin

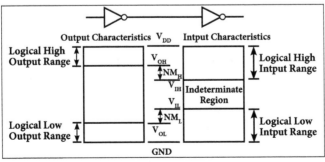

Noise Margin.

Transient Analysis,

DC analysis of V_{out} if V_{in} is constant.

Transient analysis of $V_{out(t)}$ if $V_{in(t)}$ changes.

Switching Characteristics

Switching characteristics for CMOS inverter is shown below:

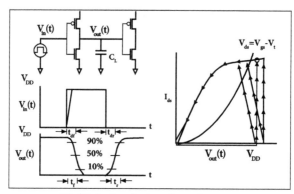

Multiple Conductor Capacitances.

- Rise time (t_r): The time for a waveform rise from 10% to 90% in its steady state value.

- Fall time (t_f): The time for a waveform fall from 90% to 10% in its steady state value.

- Delay time (t_d): The time difference between input transition (50%) and 50% output level.

  ◦ High to low delay (t_{df}).

  ◦ Low to high delay (t_{dr}).

Gate Delays

Let us consider a 3-input NAND gate as shown below:

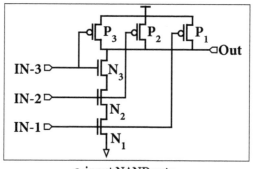

3-input NAND gate.

When pull-down path is conducting then it becomes,

$$\beta_{neff} = \frac{1}{\left(1/\beta_{n1}\right)+\left(1/\beta_{n2}\right)+\left(1/\beta_{n3}\right)}$$

For $\quad \beta_{n1} = \beta_{n2} = \beta_{n3} \Rightarrow \beta_{neff} = \dfrac{\beta_1}{3}$

Only one p-transistor has to turn on to raise the output.

Thus,

$$\beta_{peff} = \beta_p$$

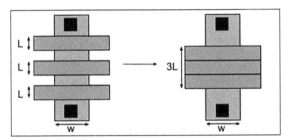

Graphical illustration of the effect of series Gate Delay.

The fall time t_f is m_{tf} $\left(t_f / m\right)$ for 'm' transistors in series (parallel). The rise time t_f for k_p transistors in series (parallel) is $k_{tr}\left(t_r / k\right)$.

Short-Circuit Power Dissipation

Even if there were no load capacitance on the output of the inverter and the parasitic are negligible, gate still dissipate switching energy.

If the input changes slowly, both the NMOS and PMOS transistors are ON. An excess power is dissipated due to the short circuit current. We are assuming the rise time of t.

Power and Ground Bounce

An example of ground bounce is shown below:

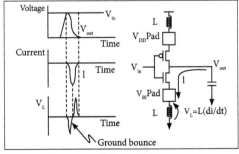

Power and Ground Bounce.

Charge Sharing

Charge, $Q = CV$

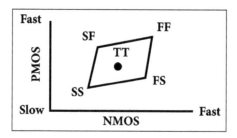

$$Q_T = C_b V_b + C_s V_s$$
$$C_T = C_b + C_s$$

$$V_R = \frac{Q_T}{C_T} = \left(C_b V_b + C_s V_s\right)/\left(C_b + C_s\right)$$

- A bus example is illustrated to explain the charge sharing phenomenon.
- A bus can be modeled as a capacitor C_b.
- An element attached to the bus can be modeled as a capacitor C_s.

Design Margining

The condition operation of a chip is influenced by some of the major factors such as:

- Operating temperature
- Supply voltage
- Process variation
- Design corners

Scaling Theory

Consider a transistor that has a channel width W and a channel length L. The main electrical characteristics change when both the dimensions are reduced by a scaling factor S > 1 such that the new transistor has size.

$$\widetilde{W} = \frac{W}{S} \quad \widetilde{L} = \frac{L}{S}$$

Gate area of the scaled transistor is given by,

$$\widetilde{A} = \frac{A}{S^2}$$

The aspect ratio of the scaled transistor is given by,

$$\frac{W}{L} = \frac{\widetilde{W}}{L}$$

The oxide capacitance is given by,

$$C_{ox} = \frac{\varepsilon_{ox}}{t_{ox}}$$

If the new transistor has a thinner oxide that is decreased as $\widetilde{t}_{ox} = \dfrac{t_{ox}}{S}$, then the scaled device has $\widetilde{C}_{ox} = SC_{ox}$.

The conductance is increased in the scaled device to,

$$\widetilde{\beta} = S\beta$$

The resistance is reduced in the scaled device to,

$$\widetilde{R} = \frac{1}{S\beta(V_{DD} - V_T)} = \frac{R}{S}$$

Assume that the supply voltage is not altered. If we can scale the voltages in device to new values of,

$$\widetilde{V}_{DD} = \frac{V_{DD}}{S} \quad \widetilde{V}_T = \frac{V_T}{S}$$

The resistance of the scaled device would be unchanged with $\widetilde{R} = R$.

The effects of scaling voltage, consider a MOS with reduced voltages of,

$$\widetilde{V}_{DS} = \frac{V_{DS}}{S} \quad \widetilde{V}_{GS} = \frac{V_{GS}}{S}$$

The current of the scaled device is given by,

$$\widetilde{I}_D = \frac{S\beta}{2}\left[\left(\frac{V_{GS}}{S} - \frac{V_T}{S}\right)\frac{V_{DS}}{S}\right] = \frac{I_D}{S}$$

The power dissipation of the scaled device is given by,

$$\widetilde{P} = \widetilde{V}_{DS}\widetilde{I}_D = \frac{V_{DS}I_D}{S^2} = \frac{P}{S^2}$$

Device Models

SPICE provides a wide variety of MOS transistor models with various tradeoffs between complexity and accuracy.

1. Level 1 Models

The SPICE Level 1 Model is closely related to model enhanced with channel length modulation and the body effect. The basic current model is given by,

$$I_{ds} = \begin{cases} 0 & V_{gs} < V_t & \text{cutoff} \\ KP \dfrac{W_{eff}}{L_{eff}}(1+LAMBDA \cdot V_{ds}) \left(V_{gs}-V_t-\dfrac{V_{ds}}{2}\right)V_{ds} & V_{ds} < V_{gs}-V_t & \text{linear} \\ \dfrac{KP}{2}\dfrac{W_{eff}}{L_{eff}}(1+LAMBDA \cdot V_{ds}) \left(V_{gs}-V_t\right)^2 & V_{ds} > V_{gs}-V_t & \text{saturation} \end{cases}$$

$$\beta = KP\dfrac{W_{eff}}{L_{eff}}$$

Where,

KP - A model parameter

W_{eff} and L_{eff} - The effective width and length

LAMBDA - Channel length modulation

The threshold voltage is modulated by the source to body voltage, V_{sb} through the body effect. For non-negative V_{sb}, the threshold voltage is given by,

$$V_t = VTO + GAMMA\left(\sqrt{PHI+V_{sb}} - \sqrt{PHI}\right)$$

Where,

VTO - Zero bias threshold voltage V_{to}

GAMMA - Body effect coefficient γ

PHI - Surface potential Φ

The Gate capacitance is calculated from the oxide thickness T_{ox}. The default gate capacitance model in HSPICE is adequate for finding the transient response of digital circuits.

Sample Level 1 Model

Model NMOS NMOS (LEVEL = 1 TOX = 40e − 10 KP = 155E − 6

LAMBDA = 0.2 VTO = 0.4 PHI = 0.93

GAMMA = 0.6 CJ = 9.8E-5

PB = 0.72 MJ = 0.36 CJSW = 2.2E − 10

PHP = 7.5 MJSW = 0.1).

Advantages:

- Easy to correlate with hand analysis.
- Too simple for modern design.

2. Level 2 and 3 Models

The SPICE Level 2 and 3 models add effects of velocity saturation, mobility degradation, Sub threshold conduction and drain induced barrier lowering. Level 2 model is based on the Grove-Frohman equations.

Level 3 models is based on empirical equations that provide similar accuracy and faster simulation times, better convergence. However, these models still do not provide good fits to the measured I-V characteristics of modern transistors.

3. BSIM Models

The Berkeley short channel IGFET Model (BSIM) is a very elaborate model that is now widely used in circuit simulation. This model have enormous number of parameters to fit the behavior of modem transistors.

BSIM versions 1, 2, 3and 4 are implemented in SPICE levels. It is quite good for digital circuit simulation except that it does not model gate leakage.

Features of the Models:

- Continuous and differentiable VI characteristics access subthreshold, linear and saturation region for good convergence.
- Sensitivity of parameters such as Vt to transistor length and width.
- Detailed threshold voltage model including body effect and drain induced barrier lowering.
- Velocity saturation, mobility degradation and other short channel effects.
- Multiple gate capacitance models.
- Diffusion capacitance and resistance models.

BSIM version 4 supports for gate leakage and other effects of very thin gates. BSIM models can be binned with different models covering different ranges of length and width specified by LMIN, LMAX, WMIN and WMAX parameters. BSIM models is not used to measure propagation delay, switching threshold, noise margins etc.

4. Diffusion Capacitance Models

The PN Junction between the source and drain diffusion body forms a reverse biased diode. Diffusion capacitance determines the parasitic delay of a gate and its depends on the area and perimeter of the diffusion.

HSPICE provides a number of methods to specify this geometry, controlled by the ACM (Area Calculation Method) parameter, which is a part of the transistor model. This model also have values for junction and sidewall diffusion capacitance.

The diffusion capacitance model is common across most device models including levels 1-3 and BSIM. HSPICE models use ACM = 0, the designer must specify the area and perimeter of the source and drain of each transistor.

Diffusion Area and Perimeter

	AS1/AD2	PS1/PD2	AD1/AS2	PD1/PS2
(a) Isolated contacted diffusion	W * 5	2 * W + 10	W * 5	2 * W + 10
(b) Shared contacted diffusion	W * 5	2 * W + 10	W * 3	W + 6
(c) Merged un-contacted diffusion	W * 5	2 * W + 10	W * 1.5	W + 3

The SPICE models also should contain parameters CJ, CJSW, PB, PHP, MJ and MJSW. Assuming the diffusion is reverse biased, area and perimeter are specified, and the diffusion capacitance between source and body is computed.

$$C_{sb} = AS \cdot CJ \cdot \left[1 + \frac{V_{sb}}{PB} \right]^{-MJ} + PS \cdot CJSW \cdot \left[1 + \frac{V_{sb}}{PHP} \right]^{-MJSW}$$

Where,

CJ → Junction capacitance.

CJSW → Side wall capacitance PB.

PHP ↛ Perimeter, MJ.

MJSW ↛ Junction gradient coefficient. BSIM3 model, the diffusion capacitance between source and body.

$$C_{sb} = AS \cdot CJ \cdot \left[1 + \frac{V_{sb}}{PB} \right]^{-MJ} + (PS - W) \cdot CJSW \cdot \left[1 + \frac{V_{sb}}{PBSW} \right]^{-MJSW}$$

$$+ W \cdot CJSWG \cdot \left[1 + \frac{V_{sb}}{PBSWG} \right]^{-MJSWG}$$

Diffusion area and perimeter used to compute the junction leakage current. This current is generally negligible compared to subthreshold leakage in modern devices.

Device Characterization

1. I-V Characteristics

The VI characteristics of NMOS and PMOS transistor shows saturation current would ideally increases quadratically with $V_{gs} - V_t$ but a linear dependence indicate that the NMOS transistor is severely velocity saturated.

The increase in saturation current with V_{ds} is caused by channel length modulation. The saturation current for a PMOS transistor is lower than the NMOS, but the device is not velocity saturated.

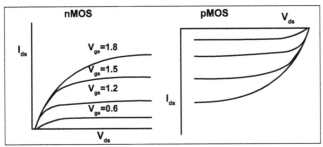

VI characteristics of NMOS and PMOS transistor.

The below figure contains:

- Straight line at low V_{gs} indicates that the current rolls off exponentially below threshold.

- Difference in subthreshold leakage at the varying drain voltage reflects the effects of drain, induced barrier lowering effectively reducing V_t at high V_{ds}.

- $I_{d\,sat}$ is measured at $V_{gs} = V_{ds} = V_{DD}$. I_{OFF} is measured at $V_{gs} = 0$.

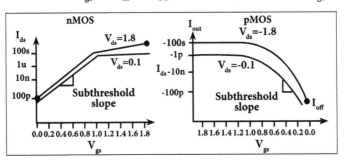

Body effect increases the threshold voltage as V_{bs}. It becomes more negative with constant as shown in the below figure.

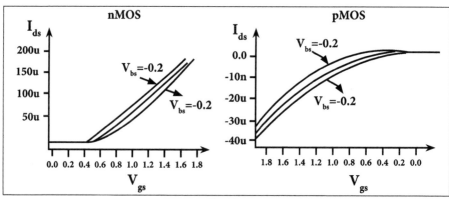

Body Effect.

2. Threshold Voltage

- In the threshold voltage V_t is defined as the value of V_{gs} at which I_{ds} becomes 0.

- In real transistor characteristics, subthreshold current continues to flow for $V_{gs} < V_t$.

- Threshold voltage varies with L, W, V_{ds} and V_{bs}.

i. Constant Current Method

This method defines threshold as gate voltage at drain current as shown in the below figure.

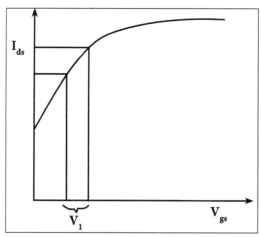

Constant Current Method.

ii. Linear Extrapolation Method

This method extrapolates the gate voltage from point of maximum slope on the $I_{ds} - V_{gs}$ characteristics.

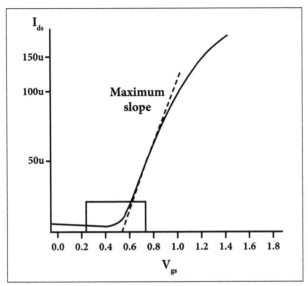

Linear Extrapolation Method.

iii. Gate Capacitance

For delay purposes, effective gate capacitance is considered to estimate gate delay for RC models. Capacitance of the data is voltage dependent and important for dynamic power consumption. Adjust the capacitance C_{delay} until the average delay from c to g equals the delay from c to d.

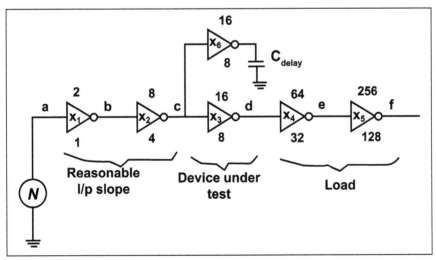

Determining effective gate capacitance.

X_6 and X_3 have the same input slope and are of the same size when they have the same delay. C_{delay} must equal the effective gate capacitance of X_4. X_1 and X_2 are used to produce a reasonable input slope on node C. X_5 is the load on X_4 to prevent node e from switching excessively fast.

Power for effective gate capacitance can be determined as shown below:

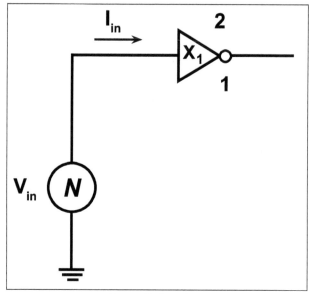

Power Estimation.

$$C_{\text{off power}} = \frac{\int I_{\text{in}}(t)\,dt}{V_{DD}}$$

iv. Parasitic Capacitance

- Parasitic capacitance associated with the source or drain of a transistor includes the gate to diffusion overlap capacitance C_{gol}, the diffusion area and perimeter capacitance C_{jb} and C_{jbsw}.

- To extract the effective parasitic capacitance for delay estimation, the circuit consists of 4 inverters and an Input/output slope.

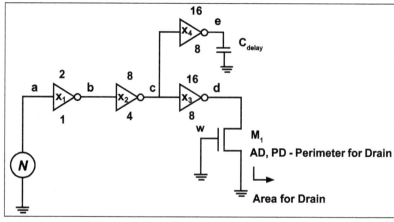

M_1 Determining effective parasitic capacitance.

- This circuit is also used to find the effective capacitance C_d.

v. Effective Resistance

According to the RC delay model, if a unit transistor has gate capacitance C, parasitic capacitance C_d and resistance R_n (for NMOS), R_p (PMOS), the rising and falling delay of a fanout of 'h' inverter with a 2:1 P/N ratio can be found.

RC Delay Model for Fanout of h-inverter

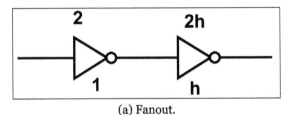

(a) Fanout.

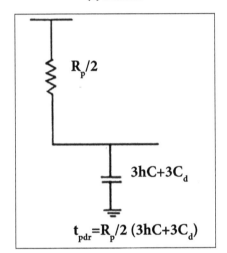

(b) $t_{pdr} = *(3hC + 3C_d)$ rising delay.

(c) $t_{pdf} = R_n(3hC + 3C_d)$ Falling delay.

Propagation Delay

Propagation delay is the time required for a signal to propagate through a gate or net. Hence, if it is cell, we can call it as "Gate or Cell Delay" or if it is net we can call it as "Net Delay".

Propagation delay of a gate or cell is a time it takes for a signal at the input pin to affect the output signal at output pin. For any gate, propagation delay is measured between 50% of input transition to the corresponding 50% of output transition.

There are 4 possibilities:

- Propagation delay between 50% of Input rising to 50% of output rising.

- Propagation delay between 50% of Input rising to 50% of output falling.

- Propagation delay between 50% of Input falling to 50% of output rising.

- Propagation delay between 50% of Input falling to 50% of output falling.

Each of these delays has different values. Maximum and minimum values of these set are very important. Maximum and minimum propagation delay values are considered for timing analysis.

For net, propagation delay is the delay between a time signal is first applied to net and time, it reaches other devices connected to that net. Propagation delay is taken as the average of rise time and fall time i.e.,

$$\tau_{pd} = \left(\tau_{phl} + \tau_{plh} \right) / 2$$

Propagation delay depends on the input transition time and the output load.

Calculation of the Propagation Delay of CMOS Inverter

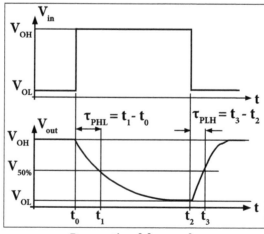

Propagation delay graph.

- Switching speed: Limited by time taken to charge and discharge, CL.

- Rise time t_r: Waveform to rise from 10% to 90% of its steady state value.

- Fall time t_f: The 90% to 10% of steady state value.

- Delay time t_d: Time difference between input transition (50%) and 50% output level.

The propagation delay t_p of gate defines, how quickly it responds to change at its inputs, it express the delay experienced by a signal when passing through a gate. It is measured between the 50% transition points of the input and output waveforms for an inverting gate.

The ζ_{pHL} defines the response time of the gate for a low to high output transition, while ζ_{pHL} refers high to low transition. The propagation delay τ_p is the average and is given by,

$$\zeta_p = \left(\zeta_{pLH} + \zeta_{pHL}\right)/2$$

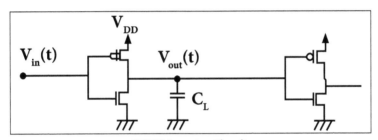

CMOS Inverter Circuit.

$$V_{50\%} = V_{OL} + \frac{V_{OH} - V_{OL}}{2} = \frac{V_{OH} + V_{OL}}{2}$$

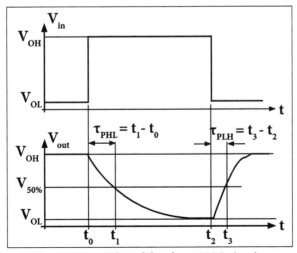

Propagation Delay of the above MOS circuit.

$$V_{90\%} = V_{OL} + 0.9(V_{OH} - V_{OL})$$

$$V_{10\%} = V_{OL} + 0.1(V_{OH} - V_{OL})$$

When $V_{in} = 0$, the capacitor C_L charges through the PMOS and when $V_{in} = 5$, the capacitor discharges through the NMOS. The capacitor current is given by,

$$C_L \frac{dV}{dt} = i_{dsn} = |i_{dsp}|$$

From this delay times can be derived as,

$$\int dt = \int \frac{C_L}{i_{ds}} dV$$

The expression for the propagation delays can be,

$$\tau_{PHL} = \frac{C_{Load} \Delta V_{HL}}{I_{avg,HL}} = \frac{C_{Load} (V_{OH} - V_{50\%})}{I_{avg,HL}} \text{ and } \tau_{PHL} = \frac{C_{Load} \Delta V_{HL}}{I_{avg,HL}} = \frac{C_{Load} (V_{50\%} - V_{OL})}{I_{avg,HL}}$$

Where $I_{avg,\,HL}$ and $I_{avg,\,LH}$ are defined as,

$$I_{avg,HL} = \frac{1}{2} \left[i_C (V_{in} = V_{OH}, V_{out} = V_{OH}) + i_C (V_{in} = V_{OL}, V_{out} = V_{50\%}) \right]$$

$$I_{avg,LH} = \frac{1}{2} \left[i_C (V_{in} = V_{OL}, V_{out} = V_{OL}) + i_C (V_{in} = V_{OL}, V_{out} = V_{50\%}) \right]$$

Rise and Fall Times

The familiar CMOS inverter with a capacity load C_L represents the load capacitance. The voltage waveform $V_{out(t)}$ is driven by a step waveform, $V_{in(t)}$ as a input.

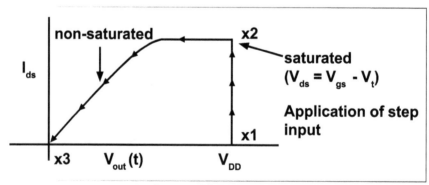

Trajectory of the n-transistor.

At n-transistor operating point an input voltage $V_{in(t)}$, changes from 0V to V_{DD}. The end device cutoff and the load capacitor are charged to V_{DD}. This is illustrated by X_1 on the characteristic curve.

Application of a step voltage $(V_{GS} = V_{DD})$ at the input of an inverter changes the operating point to X_2 and moves on the $V_{GS} = V_{DD}$ characteristic curve towards point   at the origin.

It is evident that fall time consists of two intervals:

- The t_{f1} = Period during the capacitor voltage V_{out}, drops from $0.9V_{DD}$ to $(V_{DD} - V_{tn})$.

- The t_{f2} = Period during the capacitor voltage V_{out}, drops from $(V_{DD} - V_{tn})$ to $0.1V_{DD}$.

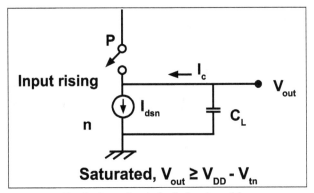

Equivalent circuit for behavior of t_{f1}.

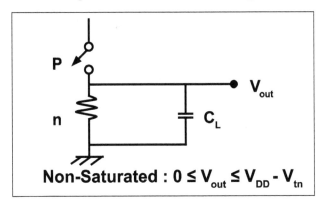

Equivalent circuit for behavior of t_{f2}.

The delay period can be derived using the general equation,

$$\int dt = \int \frac{C_L}{i_{ds}} dV$$

While in saturation,

$$I_{dsn(S\partial R)} \frac{\beta_n}{2}(V_{in} - V_{tn})^2$$

Integrating from $t = t_1$, corresponding to $V_{out} = 0.9\,V_{DD}$ to $t = t_2$ corresponding to $V_{out} = (V_{DD} - V_{tn})$ results in,

$$t_{f1} = \frac{2C_L}{\beta_n (V_{DD} - V_{tn})^2} \int_{V_{DD} - V_{tn}}^{0.9V_{DD}} dV_{out}$$

$$= \frac{2C_L (V_{DD} - 0.1V_{tn})}{\beta_n (V_{DD} - V_{tn})^2}$$

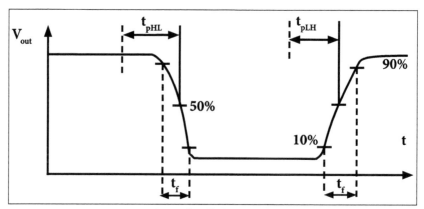

Rise and Fall time graph.

When the n-device begins to operate in linear region, discharge current is no longer constant. The time tf1 taken to discharge the capacitor voltage from $(V_{DD} - V_{tn})$ to $0.1V_{DD}$ in linear region,

$$I_{dsn(linear)} = -\beta_n \left[(V_{DD} - V_{tn}) V_{out} - V_{out}^2 / 2 \right]$$

$$t_{f2} = \frac{C_L}{\beta_n (V_{DD} - V_{tn})^2} \int_{V_{DD} - V_{tn}}^{0.9V_{DD}} \frac{dV_{out}}{\dfrac{V_{out}^2}{2(V_{DD} - V_{tn})} - V_{out}} = \frac{C_L}{\beta_n (V_{DD} - V_{tn})} \ln \left(\frac{19V_{DD} - 20V_{tn}}{V_{DD}} \right)$$

$$= \frac{C_L}{\beta_n V_{DD}(1-n)} \ln(19 - 20n)$$

Where,

$$n = \frac{V_{tn}}{V_{DD}}$$

The complete term for the fall time is given by,

$$t_f = t_{f1} + t_{f2} = \frac{2C}{\beta_n V_{DD}(1-n)} \left[\frac{(n \quad 0.1)}{1-n} + \frac{1}{2} \ln(19 - 20n) \right]$$

The fall time t_f can be approximated as,

$$t_f \approx k_n \frac{C_L}{\beta_n V_{DD}} k_n = 3 \sim 4 \text{ For } V_{DD} = 3 \sim 5V \text{ and } V_{tn} = 0.5 \sim 1V$$

From this expression, the delay is directly proportional to load capacitance. To achieve high speed circuit, minimize the load capacitance by gate. It is inversely proportional to the supply voltage i.e., the supply voltage is raised, delay time is reduced.

The delay is proportional to the β_n transistor, so increasing the width of transistor decreases the delay. Due to the symmetry of CMOS circuit the rise time can be obtained for equal size of n and p transistors (where $\beta_n = 2\beta_p$) $t_f = t_r$.

The fall time is faster than the rise time, due to different carrier mobilities associated with the p and n devices of $t_f = t_r$ if we need to make $\beta_n / \beta_p = 1$. This implies that the channel width for p-device must be increased 2 to 3 times than that of the n-device. The propagation delay is calculated as,

$$\tau_{PLH} = \frac{C_L}{k_P \left(V_{DD} - |V_{Top}|\right)} \left[\frac{2|V_{Top}|}{\left(V_{DD} - |V_{Top}|\right)} + \frac{1}{2} \ln \left(\frac{4\left(\left(V_{DD} - |V_{Top}|\right)\right)}{V_{DD}} - 1 \right) \right]$$

$$\tau_{PLH} = \frac{C_L}{k_n \left(V_{DD} - |V_{Ton}|\right)} \left[\frac{2|V_{Ton}|}{\left(V_{DD} - |V_{Ton}|\right)} + \ln \left(\frac{4\left(\left(V_{DD} - |V_{Ton}|\right)\right)}{V_{DD}} - 1 \right) \right]$$

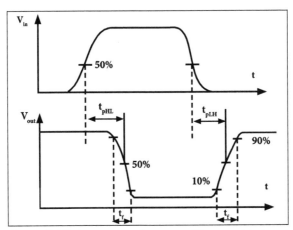

Rise and Fall time graph of output with respect to input.

If we consider the rise time and fall time of the input signal,

$$\tau_{PLH(actual)} = \sqrt{\left(\tau_{PLH}\right)^2 + \left(t_f / 2\right)^2}$$

$$\tau_{PHL(actual)} = \sqrt{\left(\tau_{PLH}\right)^2 + \left(t_f / 2\right)^2}$$

These are the 'rms' values for the propagation delays.

1.2   Complementary CMOS

The term Complementary Metal-Oxide-Semiconductor (CMOS), refers to the device technology for fabricating integrated circuits using both n- and p-channel MOSFET's. Nowadays, CMOS is the major technology in manufacturing digital IC's and is widely used in microprocessors, memories and other digital IC's. The input to a CMOS circuit is usually given to the gate of the input MOS transistor. The gate offers a very high resistance because it is isolated from the channel by an oxide layer.

The current flowing through a CMOS input is virtually zero and the device is operated mainly by the voltage applied to the gate, which controls the conductivity of the device channel. The low input currents needed by a CMOS circuit results in lower power consumption, which is the major advantage of CMOS over TTL.

In fact, power consumption in a CMOS circuit occurs only when it is switching between logic levels. Moreover, CMOS circuits are simple and cheap to fabricate resulting in high packing density than their bipolar counterparts. CMOS circuits are quite vulnerable to ESD damage, mainly by gate oxide punch through from high ESD voltages. Therefore, proper handling of CMOS IC's is needed to prevent ESD damage and generally, these devices are equipped with protection circuits.

Characteristics of CMOS logic

- Dissipates low power: The power dissipation depends on the power supply voltage, output load, frequency and input rise time. At 1 MHz and 50pF load, the power dissipation is 10nW per gate.

- Short propagation delays: Depending on the power supply, propagation delays are generally around 25nS to 50nS.

- Rise and fall times are controlled: Usually the rise and falls are ramps instead of the step functions and they are 20 to 40% longer than the propagation delays.

- Noise immunity approaches is 50% or 45% of the full logic swing.

- The logic signal level will be essentially equal to the power supplied since the input impedance is too high.

- Voltage levels ranges from 0 to V_{DD} where V_{DD} is the supply voltage.

- A low level is between 0 and $1/3$ V_{DD} while a high level is between the $2/3$ V_{DD} and V_{DD}.

NMOS Transistor

A NMOS transistor is a majority carrier device, in which the channel of current passes between the source and drain is modulated by a voltage applied to the gate.

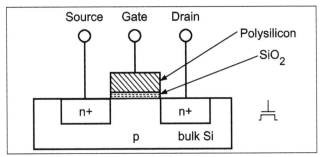

NMOS Transistor.

NMOS Operation

The terminal is commonly tied to ground (0V), when the gate is at low voltage. The low voltage is due to the following:

- P-type body is at low voltage.

- Source body and drain body diodes are OFF.

- If current flow stopped, then transistor is OFF.

Cutoff	Linear	Saturated
$V_{gsn} < V_{tn}$	$V_{gsn} > V_{tn}$	$V_{gsn} > V_{tn}$
	$V_{dsn} < V_{gsn} - V_{tn}$	$V_{dsn} > V_{gsn} - V_{tn}$

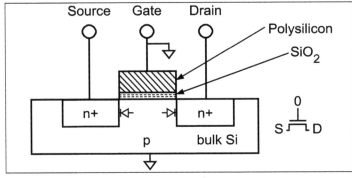

NMOS Operation 1.

$$V_{gsn} = V_{in}$$
$$V_{dsn} = V_{out}$$

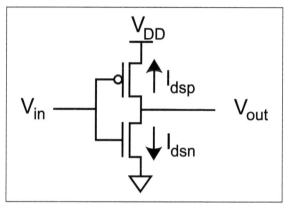

NMOS Operation 2.

Cutoff	Linear	Saturated
$V_{gsn} < V_{tn}$	$V_{gsn} > V_{tn}$	$V_{gsn} > V_{tn}$
$V_{in} < V_{tn}$	$V_{in} > V_{tn}$	$V_{out} > V_{in} - V_{tn}$
	$V_{dsn} < V_{gsn} - V_{tn}$	$V_{out} > V_{in} - V_{tn}$
	$V_{out} < V_{in} - V_{tn}$	$V_{dsn} > V_{gsn} - V_{tn}$

$V_{gsn} = V_{in}$

$V_{dsn} = V_{out}$

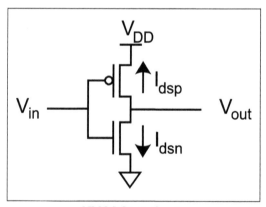

NMOS Operation 3.

When the gate is at a high voltage due to the following:

- Positive charge of gate on MOS capacitor.

- Negative charge attracted to system.

- Inverts a channel under gate to n-type.

- Now current can flow through n-type silicon from source channel to drain, then transistor is ON.

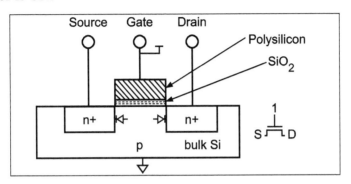

PMOS Transistor

PMOS is similar to NMOS but doping and voltage are reversed.

- The terminal or body tied to high voltage (V_{DD}).

- Gate high: transistor OFF.

- Gate low: transistor ON.

- Bubble indicates inverted behavior.

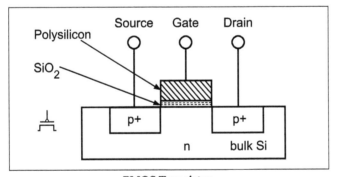

PMOS Transistor.

PMOS Operation

Cutoff	Linear	Saturated
$V_{gsp} > V_{tp}$	$V_{gsp} < V_{tp}$	$V_{gsp} < V_{tp}$
	$V_{dsp} > V_{gsp} - V_{tp}$	$V_{dsp} < V_{gsp} - V_{tp}$

$$V_{gsp} = V_{in} - V_{DD} \qquad V_{tp} < 0$$
$$V_{dsp} = V_{out} - V_{DD}$$

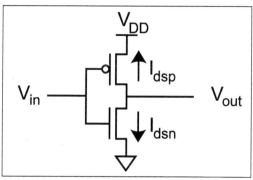

PMOS Operation 1.

Cutoff	Linear	Saturated
$V_{gsp} > V_{tp}$	$V_{gsp} < V_{tp}$	$V_{gsp} < V_{tp}$
$V_{in} > V_{DD} + V_{tp}$	$V_{in} < V_{DD} + V_{tp}$	$V_{in} < V_{DD} + V_{tp}$
	$V_{dsp} > V_{gsp} - V_{tp}$	$V_{dsp} < V_{gsp} - V_{tp}$
	$V_{out} > V_{in} - V_{tp}$	$V_{out} < V_{in} - V_{tp}$

$$V_{gsp} = V_{in} - V_{DD} \qquad\qquad V_{tp} < 0$$
$$V_{dsp} = V_{out} - V_{DD}$$

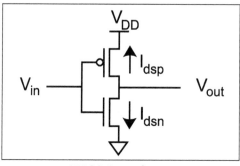

PMOS Operation 2.

I-V Characteristics make PMOS wider than NMOS such that $\beta_n = \beta_p$.

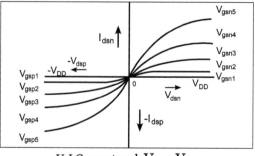

V-I Current and V_{out}, V_{in}.

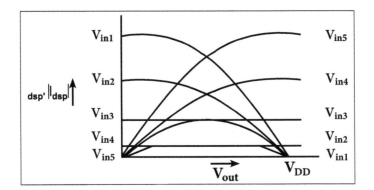

Power Supply Voltage

- GND = 0V.

- V_{DD} = 5V.

- V_{DD} has decreased in modern processes namely.

- High V_{DD} would damage modern tiny transistors.

- Lower V_{DD} saves power.

- V_{DD} = 3.3, 2.5, 1.8, 1.5, 1.2, 1.0, ...

Transistors as Switches

- We can view MOS transistors as electrically controlled switches.

- Voltage at gate controls path from source to drain.

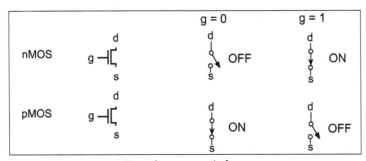

Transistors as switches.

Complementary CMOS Inverter

The D.C. transfer-characteristics relate the output voltage to the input voltage, assuming that the input voltage changes slowly so that the internal capacitance can charge or discharge fully.

The complementary CMOS inverter is shown in the below figure. Its D.C. transfer function is a curve between V_{out} Vs V_{in}.

Let,

V_{tn} be the threshold voltage of n-channel device.

V_{tp} be the threshold voltage of p-channel device (negative).

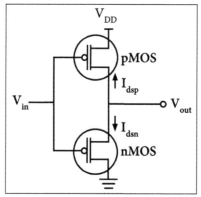

Complementary CMOS inverter.

The below table summarizes the relationship between voltages for the three regions of operation of a CMOS inverter.

Table: Regions of operations of p and n transistors.

		Cut-off	Linear	Saturated
NMOS		$V_{gsn} < V_{tn}$	$V_{gsn} > V_{tn}$	$V_{gsn} > V_{tn}$
		$V_{in} < V_{tn}$	$V_{in} > V_{tn}$	$V_{in} > V_{tn}$
			$V_{dsn} < V_{gsn} - V_{tn}$	$V_{dsn} > V_{gsn} - V_{tn}$
			$V_{out} < V_{in} - V_{tn}$	$V_{out} > V_{in} - V_{tn}$
PMOS		$V_{gsp} < V_{tp}$	$V_{gsp} < V_{tp}$	$V_{gsp} > V_{tp}$
		$V_{in} > V_{tp} + V_{DD}$	$V_{in} < V_{tp} + V_{DD}$	$V_{in} < V_{tp} + V_{DD}$
			$V_{dsp} > V_{gsp} - V_{tp}$	$V_{dsp} < V_{gsp} - V_{tp}$
			$V_{out} > V_{in} - V_{tp}$	$V_{out} < V_{in} - V_{tp}$

Since the source of NMOS transistor is grounded i.e., $V_{gsn} = V_{in}$ and $V_{dsn} = V_{out}$. As the source of PMOS transistor is connected to VDD i.e., $V_{gsp} = V_{in} - V_{DD}$ and $V_{dsp} = V_{out} - V_{DD}$.

Transfer characteristics can be found analytically or by simulation. Some of the important assumptions are as follows:

$$I_{dsn} = \left| I_{dsp} \right|$$

$$V_{tp} = -V_{tn}$$

$$\beta_n = \beta_p$$

Two important equations for MOS transistors are expressed as,

$$I_{DS} = \mu C_o \frac{W}{L} \left(V_{GS} - V_t - \frac{V_{DS}}{2} \right) V_{DS}$$

$$I_{DS} = \frac{\beta}{2} \left(V_{GS} - V_t \right)^2$$

A plot of I_{dsn} and I_{dsp} in terms of V_{dsn} and V_{dsp} for various values of V_{gn} and V_{gsp} is as shown in the below figure:

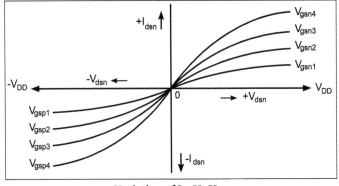

Variation of I_{DS} Vs V_{DS}.

A same plot of I_{dsn} and $\left| I_{dsp} \right|$ in terms of V_{out} for various values of V_{in} is shown in the below figure:

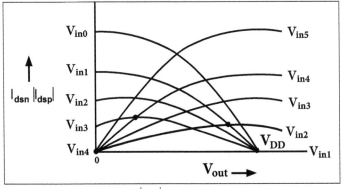

I_{dsn} and $\left| I_{dsp} \right|$ in terms of V_{out}.

The possible operating points are marked with dots. These are the values of V_{out} where $I_{dsn} = |I_{dsp}|$ for various values of V_{in}. The inverter D.C. transfer characteristics $(V_{out}$ Vs $V_{in})$ is shown in the below figure with possible-operating points.

The CMOS voltage transfer characteristic is divided into five regions. The switching point is at $V_{DD}/2$.

Region 1 $\therefore V_{in} < V_{tON}$

In this region, the NMOS transistor is in cut-off $(I_{dsn} = 0)$ and PMOS transistor is in linear region. The drain to source current I_{dsp} for p-device is zero.

$$\therefore V_{dsp} = V_{out} - V_{DD}$$

The output voltage is given by,

$$V_{out} = V_{DD} \quad (\because V_{dsp} = 0)$$

Region 2: $V_{tON} < V_{in} < V_{DD}/2$

In this region, PMOS transistor operates in the non-saturated region while NMOS transistor operates in saturation. The output voltage in this region is given by,

$$= \left(V_{in} - V_{tp}\right) + \sqrt{\left(V_{in} - V_{tp}\right) - 2\left(V_{in} - \frac{DD}{} - V_{tp}\right)V_{DD} - \frac{n}{}\left(V_{in} - V_t\right)}$$

$$V_{out} \approx V_{OH}$$

where,

β_n and β_p are the gain factors for NMOS and PMOS respectively.

(β_n and β_p are also known as K_n and K_p).

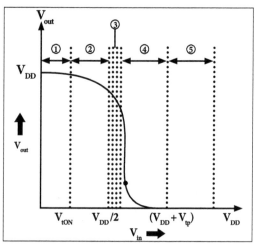

(a) CMOS voltage transfer characteristics.

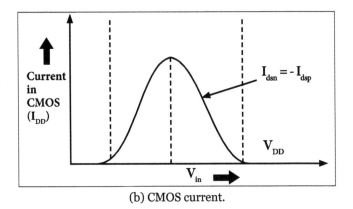

(b) CMOS current.

Region 3: $V_{in} < V_{DD}/2$

In this region, both the PMOS and NMOS transistors are in saturation region. Output voltage is given by,

$$V_{out} = V_{in}$$

This region defines the gain of CMOS inverter when used as a small signal amplifier.

Region 4: $V_{DD}/2 < V_{in} \leq (V_{DD} + V_{tp})$

In this region, PMOS transistor is in saturation region and NMOS transistor is in non-saturation region. Output voltage is given by,

$$_{out} = (V_{in} - V_{t\ ON}) - \left[(V_{in} - V_{t\ ON}) - \frac{}{} (V_{in} - V_{DD} - V_{tp}) \right]$$

$$V_{out} = V_{OL}$$

Region 5: $V_{in} \geq (V_{DD} + V_{tp})$

In this region, PMOS transistor is in cut-off region and NMOS transistor is in linear mode. Output voltage is given by,

$$V_{out} = 0$$

Table: Operating regions of PMOS and NMOS transistors and its corresponding V_{in} and V_{out}.

Regions	V_{in}	Operating regions		V_{out}
		PMOS	NMOS	
	$< V_{tON}$	Linear	Cut-off	$V_{OH} = \dfrac{V_{DD}}{2}$
	$V_{tON} \leq V_{in} < \dfrac{V_{DD}}{2}$	Linear	Saturation	$V_{OH} > \dfrac{V_{DD}}{2}$

				o(Drops sharply)
	$V_{th} = \dfrac{V_{DD}}{2}$	Saturation	Saturation	
	$\dfrac{V_{DD}}{2} < V_{in} \le \dfrac{V_{DD}}{2} - |V_{tp}|$	Saturation	Linear	$= V_{OL}\left(< \dfrac{V_{DD}}{2}\right)$
	$> \dfrac{DD}{2} + {}_{tp}$	Cut-off	Linear	0

1.2.1 Pass Transistor Logic

Pass transistor logic uses transistors as switches to carry logic signals from node to node, instead of connecting output nodes directly to VDD or ground. If a single transistor is a switch between two nodes, then voltage degradation equal to V for the high or low logic level is obtained, depending on the NMOS or PMOS transistor type.

CMOS transmission gates avoid these weak logic voltages of single-pass transistors at the cost of an additional transistor per transmission gate.

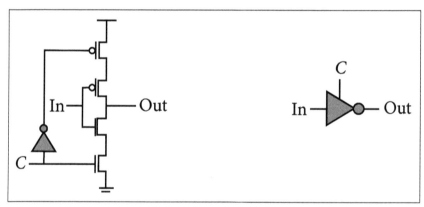

Schematic and symbol.

The transmission gate "inside" the inverter provides tri-state output.

Advantages are the low number of transistors and the reduction in associated interconnects. The drawbacks are the limited driving capability of these gates and the decreasing signal strength when cascading gates. These gates do not restore levels since their outputs are driven from the inputs and not from VDD or ground.

A typical CMOS design is the gate-level multiplexer (MUX) as shown in the figure below, for a 2-to-1 MUX. MUX selects one from a set of logic inputs to connect with the output. In the above figure, logic signal C selects either a or b to activate the output (out).

The complementary CMOS gates require 14 transistors, whereas the transmission gate design requires only six devices. Each transmission gate has two transistors plus two more to invert the control signal.

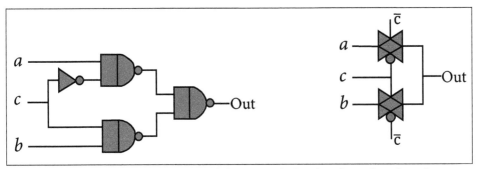

(a) Standard 2 to 1 MUX design and (b) Transmission (pass) gate based version.

Another pass gate design example is the XOR gate that produces a logic one output when only one of the inputs is logic high. If both the inputs are logic one or logic zero, then the output is zero.

The figure below shows an 8-transistor XOR gate having a tri-state buffer and transmission gate with their outputs connected. Both the gates are controlled by the same input through a complementary inverter (A-input in this case).

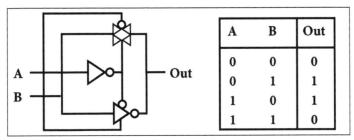

A	B	Out
0	0	0
0	1	1
1	0	1
1	1	0

8 transistor XOR gate and truth table.

The XOR gate is not a standard complementary static CMOS design since there is neither NMOS transistor network between the output and ground, nor a PMOS transistor network between the output and the power rail. The XOR standard CMOS design built previously requires fourteen transistors, whereas the design in the above figure requires only eight.

1.2.2 Transmission Gates

Transmission gates also known as pass gates represent another class of logic circuits which use TG's as basic building block. It consists of a PMOS and NMOS connected in parallel. Gate voltage applied to these gates is complementary of each other (C and Cbar).

TG's acts as bidirectional switch between two nodes A and B controlled by signal C. Gate of NMOS is connected to C and gate of PMOS is connected to Cbar. When control signal C is high i.e. V_{DD}, both transistor are on and provides a low resistance path between A and B. On the other hand, when C is low, both the transistors are turned off and provide high impedance path between A and B.

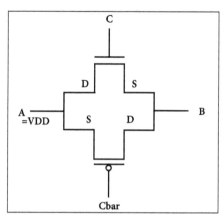

Transmission Gate.

For detailed analysis,

When input node A is connected to V_{DD} and control logic C is also high. i.e., C =1.

The output node B may be connected to capacitor. Let us say voltage at output node is V_{out}. For PMOS, source is at higher voltage than drain. For NMOS, drain is at higher voltage than Source terminal. Hence, node A will act as source terminal for PMOS and as drain terminal for NMOS.

Drain to Source and gate to source voltages for NMOS are expressed as:

$$V_{DS,n} = V_{DD} - V_{out}$$

$$V_{DG,n} = V_{DD} - V_{out}$$

For NMOS to be turned off, $V_{GS,n} < V_{th,n}$

$$V_{DD} - V_{out} < V_{th,n}$$

$$V_{DG,n} = - V_{out} > V_{DD} - V_{th,n} \text{ (cut off region)}$$

For $V_{out} < V_{DD} - V_{th,n}$

$$V_{DS,n} > V_{GS,n} - V_{th,n}$$

i.e., will operate in saturation mode

Similarly for PMOS,

$$V_{DS,p} = V_{out} - V_{DD}$$

$$V_{GS,p} = V_{DD}$$

For PMOS to be turned off $V_{GS,p} > V_{th,p}$ and threshold voltage for PMOS is -ve. Thus, always PMOS will be turned on.

For PMOS to operate in linear region, $V_{DS} > V_{GS} - V_{th,p}$

$$V_{out} - V_{DD} > -V_{DD} - V_{th,p}$$

$$V_{out} > -V_{th,p}$$

$$V_{out} > \left|V_{th,p}\right|$$

For $V_{out} \le \left|V_{th,p}\right|$, PMOS will be in saturation mode. Unlike NMOS, PMOS remain turned on regardless of output voltage V_{out}.

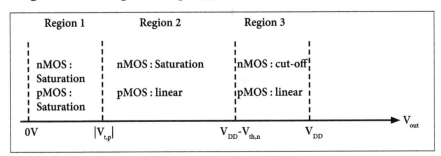

Thus, PMOS will always be turned on and as we know that, PMOS passes a strong 1 so voltage level high will be transmitted.

Similarly, when voltage at node A, $V_{in} = 0$ and $C = V_{DD}$: Node A will act as source terminal for NMOS and will act as drain for PMOS. NMOS will always be turned on hence, level 0 will also be transmitted unattenuated.

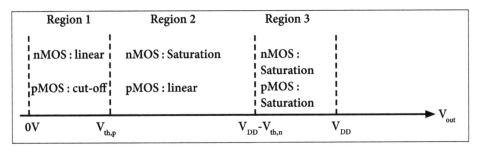

When voltage at node A, $V_{in} = V_{DD}$ and $C = 0$: Node A will act as drain terminal for NMOS and source terminal for PMOS.

$$V_{GS,n} = 0 - V_{DD} < V_{th,n} \text{ (Cut off region)}$$

Hence, NMOS will be turned off.

$$V_{GS,p} = V_{DD} - V_{DD} = 0 > V_{th,p} \text{ (Cut off region)}$$

Thus, both the transistor will remain off. Path between A and B will be an open circuit. When voltage at node A, $V_{in} = 0$ and C = 0: Node A will act as source terminal for NMOS and will act as drain for PMOS.

$$V_{GS,n} = 0 - 0 = 0 \; V_{th,n} \; \text{(Cut off region)}$$

$$V_{GS,p} = V_{DD} - V_{out}$$

$V_{GS,p}$ will be some positive voltage and threshold voltage of PMOS, $V_{th,p}$ is negative.

$$V_{GS,p} > V_{th,p} \; \text{(Cut off region)}$$

Hence, both the transistor will remain off and high impedance path exists between A and B.

1.2.3 Technology Scaling and its Impact on the Inverter Metrics

We can derive that the gate delay indeed decreases exponentially at a rate of 13%/year, or halving every five years. This rate is on course with the prediction, since S averages approximately 1.15. The delay of a 2-input NAND gate with a fanout of four has gone from tens of nanoseconds in the 1960s to a tenth of a nanosecond in the year 2000, and is projected to be a few tens of picoseconds by 2010.

Reducing power dissipation has only been a second-order priority until recently. Hence, statistics on dissipation-per-gate or design are only marginally available. Although the variation is large, even for a fixed technology, it shows the power density to increase approximately with S2. This is in correspondence with the fixed-voltage scaling scenario presented in the below table.

Table: Scaling scenarios for short-channel devices.

(S and U represent the technology and voltage scaling parameters, respectively).

Parameter	Relation	Full Scaling	General Scaling	Fixed-Voltage Scaling
Area/Device	WL	$1/S^2$	$1/S^2$	$1/S^2$
Intrinsic Delay	$R_{on}C_{gate}$	$1/S$	$1/S$	$1/S$
Intrinsic Energy	$C_{gate}V^2$	$1/S^3$	$1/SU^2$	$1/S$
Intrinsic Power	Energy/Delay	$1/S^2$	$1/U^2$	1
Power Density	P/Area	1	S^2/U^2	S^2

For more recent years, we expect a scenario more in line with the full-scaling model, which predicts a constant power density due to the accelerated supply-voltage scaling and the increased attention to power-reducing design techniques. Even under these circumstances, power dissipation-per-chip will continue to increase due to the ever larger die sizes.

The performance and power predictions produce purely "intrinsic" numbers that take only device parameters into account. It was concluded that the interconnect wires exhibit a different scaling behavior, and that wire parasitic may come to dominate the overall performance. Similarly, charging and discharging the wire capacitances may dominate the energy budget.

To get a crisper perspective, one has to construct a combined model that consider device and wire scaling models simultaneously. The impact of the wire capacitance and its scaling behavior is summarized in the below table.

Table: Scaling scenarios for wire capacitance.

Parameter	Relation	General Scaling
Wire Capacitance	WL/t	$\mathring{a}_c/S_L$
Wire Delay	$R_{on}C_{int}$	$\mathring{a}_c/S_L$
Wire Energy	$C_{int}V^2$	ε_c/S_LU^2
Wire Delay / Intrinsic Delay		$\mathring{a}_cS/S_L$
Wire Energy / Intrinsic Energy		ε_cS/S_L

Here, S and U represent the technology and voltage scaling parameters, respectively, while S_L stands for the wire-length scaling factor. ε_c Represents the impact of fringing and inter-wire capacitances.

We furthermore assume that the resistance of the driver dominates the wire resistance, which is definitely the case for short to medium-long wires. The model predicts that the interconnect-caused delay gain in importance with the scaling of technology. This impact is limited to an increase with ε_c for short wires $(S = S_L)$, but it becomes increasingly more outspoken for medium-range and long wires $(S_L < S)$.

However, the ratio of wire over intrinsic contributions will actually evolve is debatable, as it depends upon a wide range of independent parameters such as system architecture, design methodology, transistor sizing and interconnect materials.

The doomday scenario that interconnect may cause the CMOS performance to saturate in the very near future hence, may be exaggerated. Yet, it is clear that increased attention to interconnect is an absolute necessity and may change the way the next-generation circuits are designed and optimized.

1.3   Dynamic CMOS Logic and Timing Metrics

Basic Structure

A dynamic CMOS gate implements the logic with a block of transistors. The output node is connected to ground through an NMOS transistor block and a single NMOS evaluation transistor. The output node is connected to the power supply through one pre-charge PMOS transistor. A global clock drives the pre-charge and evaluation transistors.

The gate has two phases such as evaluation and pre-charge. During pre-charge, the global clock goes low, turning the PMOS transistor on and the evaluation NMOS off. The gate output goes high while the block of NMOS transistors float.

In the evaluation phase, the clock is driven high, turning the PMOS device off and the evaluation NMOS on. The input signals determine if there is a low or high impedance path from the output to ground since the global clock turns on the NMOS evaluation transistor. This design eliminates the speed degradation and power wasted by the short-circuit current of the n-channel and p-channel transistors during the transition of static complementary designs.

If the logic state determined by the inputs is a logic one (V_{DD}), then the rise time is zero. The pre-charge and evaluation transistors are designed to never conduct simultaneously. Dynamic circuits with an n-input gate use only n+2 transistors instead of the 2n devices required for the complementary CMOS static gates. Dynamic CMOS gates have a drawback.

If the global clock in the figure below is set high, then the output node could be in high-state with no electrical path to V_{DD} or ground. This exposes the node to noise fluctuations and charge sharing within the logic block, thus degrading its voltage. Also, the output load capacitor will slowly discharge due to transistor off-state leakage currents and loses its logic value. This limits the low-frequency operation of the circuit.

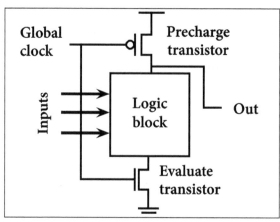

Basic structure of dynamic CMOS gate.

The gate inputs can only change during pre-charge, since charge redistribution from the output capacitor to internal nodes of the NMOS logic block may drop the output voltage when it has logic high.

Finally, dynamic gate cascading is challenging since differences in delay between logic gates may cause a slow gate to feed an erroneous logic high to the next gate. This would cause the output of the second gate to be erroneously zero. Different clocking strategies can avoid this.

Domino CMOS Logic

Domino CMOS was proposed by Krambeck et al. in 1982. It has the same structure as dynamic logic gates, but adds a static buffering CMOS inverter to its output. In some cases, there is also a weak feedback transistor to latch the internal floating node high when the output is low. This logic is the most common form of dynamic gates, achieving a 20-50% performance increase over static logic.

When the NMOS logic block discharges, the output node during evaluation, the inverter output out goes high, turning off the feedback PMOS. When out is evaluated high, then the inverter output goes low, turning on the feedback PMOS device and providing a low impedance path to VDD. This prevents the out node from floating, making it less sensitive to node voltage drift, noise and current leakage.

Domino CMOS allows logic gate cascading since all the inputs are set to zero during pre-charge, avoiding erroneous evaluation from different delays. This logic allows static operation from the feedback latching PMOS, but logic evaluation still needs two sub cycles: pre-charge and evaluation.

Domino logic uses only non-inverting gates, making it an incomplete logic family. To achieve the inverted logic, a separate inverting path running in parallel with the non-inverted one must be designed.

Multiple output domino logic (MODL) is an extension of domino logic, taking internal nodes of the logic block as signal outputs, thus saving area, power and performance. Compound domino logic is another design that limits the length of the evaluation logic to prevent charge sharing and adds other complex gates such as buffer elements (NAND, NOR, etc. instead of inverters) to obtain more area compaction. Self-resetting domino logic (SRCMOS) has each gate detect its own operating clock. Thus, reducing clock overhead and providing high performance.

NORA CMOS Logic

This design alternative to domino CMOS logic eliminates the output buffer without causing race problems between clock and data that arise when cascading dynamic gates. NORA CMOS (No-Race CMOS) avoids these race problems by cascading the alternate NMOS and PMOS blocks for logic evaluation. The cost is routing two complemented clock signals. The cascaded NORA gate structure is shown in the above figure.

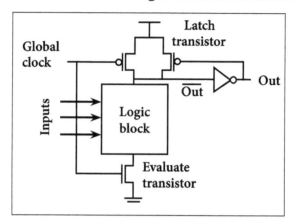

Domino CMOS logic gate with feedback transistor.

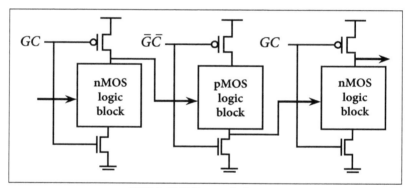

NORA CMOS cascaded gates.

When the global clock (GC) is low (GC high), the NMOS logic block output nodes are pre-charged high while the outputs of gates with PMOS logic blocks are pre-charged low. When the clock changes, gates are in the evaluate state.

Other CMOS Legit Families

Dynamic circuits have a clock distribution problem since all gates must be functionality synchronized.

Self-timed circuits are an alternative to dynamic high-performance circuits, solving the clock distribution which does not requires a global clock. This simplifies clock routing and minimizes clock skew problems related to clock distribution.

The global clock is replaced by a specific self-timed communication protocol between circuit blocks in a request-acknowledge scheme. Although more robust than dynamic circuits, self-timed logic requires a higher design effort than other families. These gates implement self-timing by using differential cascade voltage switch logic (known as DCVS) based on an extension of the domino logic.

The DCVS logic family uses two complementary logic blocks, in which each blocks are similar to the domino structure. The gate inputs must be in the true and complementary form. Since both the true output and negated output are available, they can activate a completion signal when the output is evaluated.

Since the gate itself signals when the output is available, DCVS can operate at the maximum speed of the technology, providing high performance asynchronous circuits. The major drawbacks are design complexity and increased size.

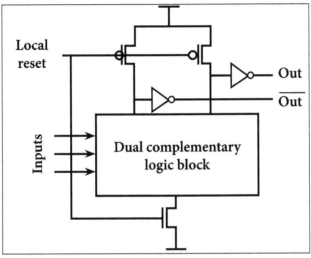

Basic DCVS logic gate.

1.3.1 CMOS Logic Design Perspectives

Power Dissipation

Static CMOS gates in older technologies were very power-efficient. In newer technologies, power is a primary design constraint. Power dissipation has skyrocketed due to the transistor scaling, chip transistor counts and clock frequencies.

Instantaneous Power

The instantaneous power $P(t)$ drawn from the power supply is proportional to the supply current $i_{DD}(t)$ and the supply voltage V_{DD}.

$$P(t) = i_{DD}(t)\, V_{DD}$$

Energy

The energy consumed over the time interval T is the integral of $P(t)$.

$$E = \int_0^T i_{DD}(t) V_{DD} dt$$

Average Power

The average power over this interval is given by,

$$P_{avg} = \frac{E}{T} = \frac{1}{T}\int_0^T i_{DD}(t)V_{DD}dt$$

CMOS Power Dissipation

Power dissipation in CMOS circuits comes from two components such as:

- Static dissipation due to:

  ○ Tunneling current through gate oxide.

  ○ Subthreshold conduction through OFF transistors.

  ○ Contention current in ratioed circuits.

  ○ Leakage through reverse-biased diodes.

  ○ Charging and discharging of load capacitances.

  ○ Short circuit current while both PMOS and NMOS networks are partially ON.

- Dynamic dissipation due to:

$$P_{total} = P_{static} + P_{dynamic}$$

Below 130nm, static power is rapidly becoming a primary design issue. Eventually, static power dissipation may become comparable to dynamic power, ratioed circuits (e.g. pseudo NMOS) have more static dissipation.

Dynamic Power Dissipation

Primary source of dynamic dissipation is charging of the load capacitance. Suppose load C is switched between V_{DD} and GND at average frequency f_{sw} over time T, load is charged and discharged T_{fsw} times. In one complete charge/discharge cycle, a total charge of $Q = CV_{DD}$ is transferred between V_{DD} and GND.

The average dynamic power dissipation is given by,

$$P_{dynamic} = \frac{1}{T}\int_0^T i_{DD}(t)V_{DD}dt = \frac{V_{DD}}{T}\int_0^T i_{DD}(t)dt$$

Taking the integral of the current over interval T as the total charge delivered during time T,

$$P_{dynamic} = \frac{V_{DD}}{T}[Tf_{sw}CV_{DD}] = CV_{DD}^2 f_{sw}$$

As not all gates switch every clock cycle the above quantity is multiplied by α. $\alpha = 1$ for clock, for data maximum is $\alpha = 0.5$, and empirically static CMOS has $\alpha = 0.1$

Also, due to non-zero input rise and fall times (slew), both NMOS and PMOS will be ON causes short circuit current that depends on input slew and output capacitance.

Total Power Dissipation

Total power dissipation is the sum of the static and dynamic dissipation components. Dynamic dissipation has historically been far greater than static power when systems are active and hence, static power is often ignored, although this will change as gate and subthreshold leakage increase.

Power dissipation has become extremely important to VLSI designers. For high-performance systems such as workstations and servers, dynamic power consumption per chip is often limited to about 150W by the amount of heat that can be managed with air cooled systems and cost-effective heat sinks.

This number increases slowly with advances in the heat-sink technology and can be increased significantly with the expensive liquid cooling, but does not kept pace with the growing power demands of systems. Hence, its performance may be limited by the inability to cool huge systems with power-hungry circuits operating at high speeds.

For battery-based systems such as laptops, cell phones and PDA's, power consumption sets the battery life of the product. All of the switching activity may be stopped in an idle or "sleep" mode. Hence, in addition to dynamic power while active, static power consumption may limit the battery life while idle.

Dynamic Power Reduction

If a process is selected with sufficiently high threshold voltages and oxide thicknesses, static dissipation is small and dynamic dissipation usually dominates while the chip is active. Equation below shows that the dynamic power is reduced by decreasing the activity factors, the switching capacitance, the power supply or the operating frequency.

$$P_{dynamic} = \alpha C V_{DD}^2 f$$

Activity factor reduction is very important. Static logic has an inherently low activity factor. Clocked nodes such as clock network and clock input to registers have an activity factor of 1 and are very power-hungry. Dynamic circuit families have clocked nodes and a high internal activity factor, so they are also costly in power.

Clock gating can be used to stop portions of the chip that are idle. For example, a floating point unit can be turned off when executing the integer code and a second level cache can be idled if the data is found in the primary cache. A large fraction of power is dissipated by the clock network itself, so entire portions of the clock network can be turned off where possible. The chip can also sense die temperature and cut back activity if the temperature becomes too high.

A drawback of activity factor reduction is that if the system transitions rapidly from an idle mode with little switching to a fully active mode, large d_i/d_t spike will occur. This leads to inductive noise in the power supply network. Some systems throttle execution, limiting the number of functional units that go from idle to active in each cycle.

Device-switching capacitance is reduced by choosing small transistors. Minimum-size gates can be used on non-critical paths. Although logical effort finds that the best stage effort is about 4, using a larger stage effort increases delay only slightly and greatly reduces transistor sizes.

For example, buffers driving I/O pads or long wires may use a stage effort of 8-12 to reduce the buffer size. Interconnect switching capacitance is most effectively reduced through careful floor planning, placing communicating units near each other to reduce wire lengths.

Voltage has a quadratic effect on the dynamic power. Therefore, choosing a lower power supply significantly reduces power consumption. As many transistors are operating in a velocity-saturated regime, the lower power supply may not reduce performance as much as first-order models predict.

Voltage can be adjusted based on the operating mode. A laptop processor may operate at high voltage and high speed when plugged into an AC adapter, but at lower voltage and speed when on battery power.

If the frequency and voltage scale down in proportion, a cubic reduction in power is achieved. For example, 1. The laptop processor may scale back to 2/3 frequency and voltage to save 70% in power when unplugged. Frequency can also be traded for power. 2. In a digital signal processing system primarily concerned with throughput, two multipliers running at half speed can replace a single multiplier at full speed.

At first, this may not appear to be a good idea because it maintains constant power and performance while doubling area. However, if the power supply can also be reduced because the frequency requirement is lowered, overall power consumption goes down.

Commonly used metrics in low-power design are power, the power-delay product and the energy-delay product. Power alone is a questionable metric because it can be reduced simply by computing more slowly. The power-delay (i.e., energy) product is also suspect, because the energy can be reduced by computing more slowly at a lower supply voltage. The energy-delay product (i.e., power * delay^2) is less prone to such gaming.

Overall, the energy-delay product measured in Performance^2/Watt normalized for process only varies by about a factor of two across a wide range of general-purpose microprocessor architectures. This suggests that as long as wasteful practices are avoided, there is little we can do to the general-purpose processors except trade the energy consumed by a computation against the delay of the computation.

The big power gains are to be made not through tweaking of circuits but by reconsidering algorithms. For example, the Fast Fourier Transform requires fewer arithmetic operations and hence less power than a Discrete Fourier Transform. Signal-processing systems using data paths are hardwired to a particular operation consume far less power than general-purpose processors delivering the same performance because the data paths eliminate unnecessary control units.

Static Power Reduction

Static power reduction involves minimizing I_{static}. Some circuit techniques such as analog current sources and pseudo-NMOS gates intentionally draw static power. They can be turned off when they are not needed.

Recall that the subthreshold leakage current for $V_{gs} < V_t$ is given by,

$$I_{ds} = I_{dro}e^{\frac{V_{gs}-V_t}{nv_T}}\left[1-e^{\frac{-V_{ds}}{v_T}}\right]$$

$$V_t = V_{to} - \eta V_{ds} + \gamma\left(\sqrt{\phi_s + V_{sb}} - \sqrt{\phi_s}\right)$$

Where the η term describes drain-induced barrier lowering and the γ term describes the body effect. For any appreciable V_{ds}, the term in brackets approaches unity and can be discarded.

The remaining term can be reduced by increasing the threshold voltage V_{to}, reducing V_{gs}, V_{ds}, increasing V_{sb} or lowering the temperature.

Subthreshold leakage power is already a major problem for battery-powered designs in the 180nm generation and will be growing exponentially as power supplies and threshold voltages are scaled down in future processes. Many low-power systems need high performance while active and low leakage while idle.

The high-performance requirement entails relatively low thresholds, which contribute excessive leakage current in the idle mode. Selective application of multiple threshold voltages can maintain performance on critical paths with low-V_t transistors while reducing leakage on other paths with high-V_t transistors.

In low-power battery-operated devices, leakage specifications may be given at 40°C rather than 110°C because battery life is most important in the range of normal ambient temperatures. Another way to control leakage is through body voltage using body effect.

For example, low-V_t devices can be used and a reverse body bias (RBB) can be applied during idle mode to reduce leakage. Alternatively, higher V_t devices can be used and then a forward body bias (FBB) can be applied during active mode to increase performance.

The threshold voltages may vary from one die to another on account of manufacturing variations. An adaptive body bias (ABB) can compensate and achieve more uniform transistor performance despite the variations. In any case, the body bias should be kept to less than about 0.5 V.

Too much of reverse body bias leads to greater junction leakage through a mechanism called band-to-band tunneling, while too much forward body bias leads to substantial current through the body to source diodes.

Applying a body bias requires additional power supply rails to distribute the substrate and well voltages. For example, a RBB scheme for a 1.8 V n-well process could bias the p-type substrate at $V_{BBn} = 0.4$ V and the n-well at $V_{BBp} = 2.2$ V. The figure below shows a schematic and cross-section of an inverter using body bias.

In an n-well process, all NMOS transistors share the same p substrate and must use the same V_{BBn}. In a triple-well process, groups of transistors can use different p-wells isolated from the substrate and thus can use different body biases. The well and substrate carry little current, so the bias voltages are relatively easy to generate and distribute.

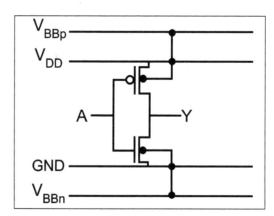

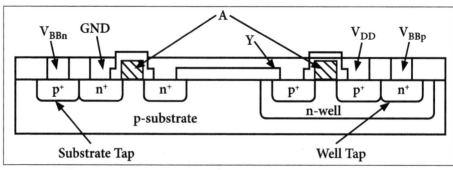

Body Bias.

Alternatively, the source voltage can be raised in sleep mode. This has the double benefit of reducing V_{ds} as well as increasing V_{sb}. However, the source does carry the significant current, thus generating a stable and adjustable source voltage rail is challenging.

Reducing V_{DD} in standby mode reduces the drain-induced barrier lowering contribution to leakage. It also decreases gate leakage in processes where that component is important. The supply should be maintained at a high enough level to preserve the state of the system.

Yet, another method of reducing idle leakage current in low-power systems is to turn off the power supply entirely. This could be done externally with the voltage regulator or internally with the series transistor.

Multiple Threshold CMOS circuits (MTCMOS) uses low-V_t transistors for computation and a high-V_t transistor as a switch to disconnect the power supply during idle mode. The high-V_t device is connected between the true V_{DD} and the virtual V_{DDV} rails connected to the logic gates. The extra transistor increases the impedance between true and virtual power supply, causing greater power supply noise and gate delay.

Bypass capacitance between V_{DDV} and GND stabilizes the supply, but the capacitance is discharged each time V_{DDV} is disconnected, contributing to the power consumption. Even using a very wide high-V_t transistor, MTCMOS is only suited to systems with small power demands. The PMOS body should be tied to V_{DD} so both V_{DD} and V_{DDV} lines must be routed to all cells. MTCMOS uses carefully designed registers connected to the true supply rails to retain state during idle mode.

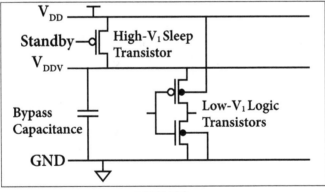

MTCMOS.

The leakage through two series OFF transistors is much lower than that of a single transistor because of the stack effect. In the below figure (a), the single transistor has a relatively low threshold because of drain-induced bather lowering from the high drain voltage. In the figure (b), node x rises to about 100mV. The threshold on the bottom transistor is higher because of the small drain voltage. The top transistor also turns off harder because of the negative V_{gs} and the body effect. The net result is that I_2 maybe 10-20 times smaller than I_1.

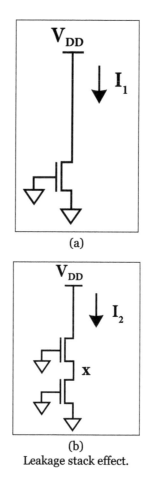

(a)

(b)

Leakage stack effect.

Low-power systems can take advantage of this stack effect to put gates with series transistors into a low-leakage sleep mode by applying an input pattern to turn off both transistors. Silicon on Insulator (SOI) circuits are attractive for low-leakage designs because they have a sharper sub-threshold current roll-off.

1.4   Timing Metrics for Sequential Circuits

Physically implemented combinational circuits (NAND or NOR gates) exhibit certain timing characteristics.

A "0" or "1" applied at the input to a combinational circuit does not result in an instantaneous change at the output because of various electrical constraints. Input-to-output delay in the combinational circuits can be expressed with two parameters such as propagation delay, t_{pd} and contamination delay t_{cd}.

Propagation delay (t_{pd}): The amount of time needed for a change in a logic input to result in a permanent change at an output, i.e., the combinational logic will not show any further output changes in response to an input change after time t_{pd} units.

Contamination delay (t_{cd}): The amount of time needed for a change in a logic input to result in an initial change at an output, i.e., the combinational logic is guaranteed not to show any output change in response to an input change before t_{cd} time units have passed.

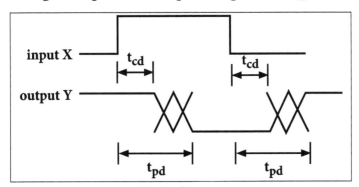

Timing diagram.

Combinational propagation delays are additive and thus the propagation delay of a larger combinational circuit can be determined by adding the propagation delays of each of the circuit components along the longest path. In contrast, finding the contamination delay of the circuit requires identifying the shortest path of contamination delays from input to output and adding the delay values along this path.

Timing Parameters for Sequential Logic

When the sequential circuits are physically implemented, they exhibits certain timing characteristic which is unlike the combinational circuits, which are specified in relation to the clock input.

Latch vs Flip-Flop

A latch is level-sensitive while a flip-flop is edge triggered. A latch stores when the clock level is low and is transparent when the level is high. A flip-flop stores when the clock rises and is mostly never transparent. Since the flip-flops only changes the values in response to the change in the clock value, timing parameters can be specified in relation to the rising or falling clock edge.

The following parameters specify the sequential circuit behavior. Note that these are all for positive edge-triggered flip-flops unless otherwise specified, but are easily applied to negative edge triggered flip-flops as well.

- Propagation delay $(t_{clk}-q)$: The amount of time needed for a change in the flip-flop clock input D to result in a change at the flip-flop output Q. When the clock edge arrives, D input value is transferred to the output Q. After time $t_{clk}-q$, the output is guaranteed not to change the value again until another clock edge trigger arrives.

- Contamination delay (t_{cd}): This value indicates the amount of time needed for a change in the flip-flop clock input to result in the initial change at the flip-flop output Q. The output of the flip-flop maintains its initial value until time t_{cd} has passed and is guaranteed not to show any output change in response to an input change until after t_{cd} has passed. Delays can be different for both the rising and falling transitions.

- Setup time (t_{su}): The amount of time before the clock edge that data input D must be stable, the rising clock edge arrives.

- Hold time (t_{hold}): This indicates the amount of time after the clock edge arrives that data input D must be held stable in order to latch the correct value. Hold time is always measured from the rising clock edge to a point after the clock edge.

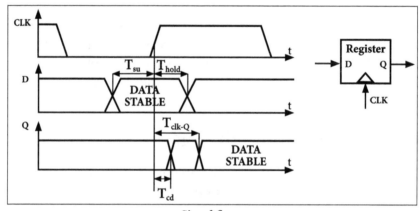

Signal flow.

Setup and hold times are restrictions that a flip-flop places on combinational or sequential circuitry that drives a flip-flop D input. The circuit has to be designed so that the D input signal arrives at least t_{su} time units before the clock edge and does not change until at least t_{hold} time units after the clock edge. If either of these restrictions is violated for any of the flip-flops in the circuit, the circuit will not operate correctly. These restrictions limit the maximum clock frequency at which the circuit can operate.

Determining the Maximum Clock Frequency for a Sequential Circuit, most of the digital circuits contains both the combinational components (gates, muxes, adders, etc.) and sequential components (flip-flops). These components can be combined to form the sequential circuits that perform computation and store results.

By using combinational and sequential component parameters, it is possible to determine the maximum clock frequency at which a circuit will operate and generate correct results.

Delay Modeling

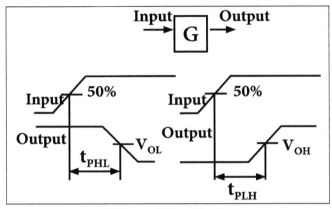

Delay modelling circuit.

Gate Propagation Delay (tPHL and tPLH)

Gate propagation delay is measured from 50% input to 50% of the output. t_{PHL} is measured from 50% of the rising edge of input voltage to 50% of the rising edge of output voltage. Similarly, t_{PLH} is measured from 50% of the falling edge of input voltage to 50% of the falling edge of output Voltage.

Clock Trees

If the number of flip-flops driven by the clock line is large, the clock rise time (also called as slew rate) will be unacceptably long. The solution to this problem is to use a clock power up tree which means adding buffers into the clock tree.

When designing the clock distribution network, the absolute delay from a central clock source to the clock elements is irrelevant, only the relative phase between two clock elements is important.

When designing clock trees in this way, first determine the number of levels, our clock tree can have. This will depend on the total number of flip-flops in the circuit and the number of fan outs that we are limited to.

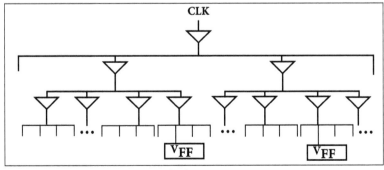

Clock Trees.

Clock Skew (δ)

The spatial variation in the arrival time of a clock transition is known as clock skew. The clock skew between two points j and k is given by $t_j - t_k$, where $t_j - t_k$ are the rising edge of the clock with respect to the reference. Clock skew is constant from cycle to cycle and does not cause clock period variation, but only phase shift.

Clock Jitter

Clock jitter refers to the temporal variation of the clock period, i.e., the clock period can expand or reduce on a cycle-by-cycle basis.

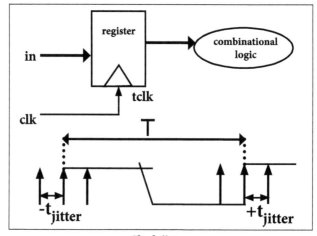

Clock jitter.

Variation of the pulse width is important for level sensitive clocking.

Positive and Negative Clock Skew

Positive Skew: Clock and data flow is in the same direction.

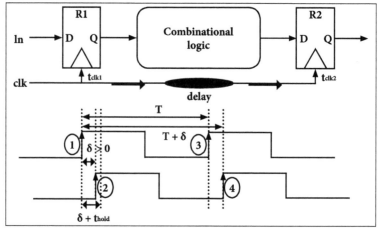

Positive Skew.

$$T: \quad T + \delta \geq t_{c-q} + t_{plogic} + t_{su} \text{ so } T \geq t_{c-q} + t_{plogic} + t_{su} - \delta$$

$$t_{hold}: \quad t_{hold} + \delta \leq tc_{dlogic} + t_{cdreg} \text{ so } t_{hold} \leq t_{cdlogic} + t_{cdreg} - \delta$$

Negative Skew: Clock and data flow is in the opposite directions.

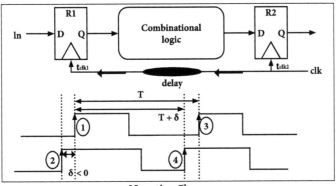

Negative Skew.

$$T: \quad T + \delta \geq t_{c-q} + t_{plogic} + t_{su} \text{ so } T \geq t_{c-q} + t_{plogic} + t_{su} - \delta$$

$$t_{hold}: \quad t_{hold} + \delta \leq tc_{dlogic} + t_{cdreg} \text{ so } t_{hold} \leq t_{cdlogic} + t_{cdreg} - \delta$$

Internally Generated Clocks

A designer should avoid internally generated clocks, wherever, possible, as they can cause functional and timing problems in the design, if not handled properly. Clocks generated with combinational logic can introduce glitches that create functional problems and the delay due to the combinational logic can lead to timing problems.

In a synchronous design, a glitch on the data inputs does not cause any issues and is automatically avoided as data is always captured on the edge of the clock and thus blocks the glitch. However, a glitch or a spike on the clock input can have significant consequences.

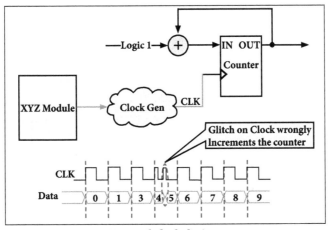

Internal clock design.

Setup and hold times may also be violated if the data input of the register is changing when a glitch reaches the clock input. Narrow glitches can violate the register's minimum pulse width requirements. Even if the design does not violate the timing requirements, the register output can change value unexpectedly and cause functional hazards elsewhere in the design.

The above figure shows the effect of using combinational logic to generate the clock on a synchronous counter. As shown in the timing diagram, due to the glitch on the clock edge, the counter increments twice in the clock cycle.

Counter example for using combinational logic as a clock:

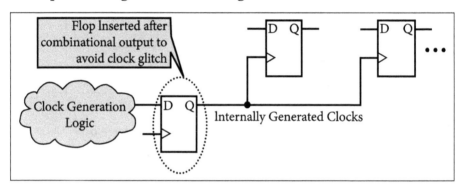

Counter example for using combinational logic as a clock. This extra counting may create functional issues in the design where instead of counting the desired count, counter counts an additional count due to the glitch on the clock.

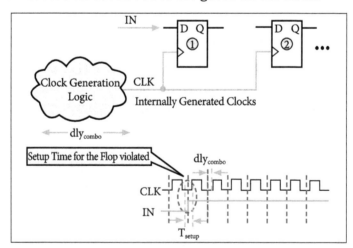

Setup Time Violated Due to Skew of Clock Path

The combinational logic used to generate the internal clock also adds delays on the clock line. In some cases, logic delay on a clock line can result in a clock skew greater than the data path length between two registers. If the clock skew is greater than the data delay, the timing parameters of the register will be violated and the design will not function correctly.

Divided Clocks

Many designs require clocks created by dividing a master clock. Design should ensure that most of the clocks should come from the PLL. PLL circuitry will avoid many of the problems which can be introduced by asynchronous clock division logic. While using logic to divide a master clock, always use synchronous counters or state machines.

In addition, the design should ensure that registers always directly generate divided clock signals. Design should never decode the outputs of a counter or a state machine to generate clock signals. This type of implementation often causes glitches and spikes.

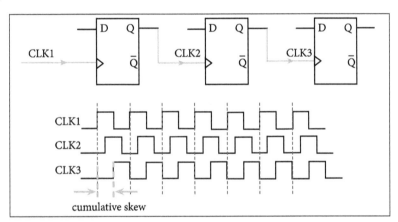

Ripple Counters

ASIC designers have often implemented the ripple counters to divide the clocks by a power of 2 because the counter uses fewer gates than their synchronous counterparts. Ripple counters uses cascaded registers, in which the output pin of each register feeds the clock pin of the register in the next stage.

This cascading can cause problems because the counter creates a ripple clock at each stage. These ripple clocks pose another set of challenges for STA and synthesis tools. We should try to avoid these types of structures to ease verification effort. Despite of all the challenges and problems with respect to Ripple counters, these are quite handy in systems which is good to reduce the peak power consumed by the logic or SoC.

Multiplexed Clocks

Clock multiplexing can be used to operate the same logic function with different clock sources. Multiplexing logic of some kind selects a clock source as shown in the below figure. For example, telecommunications applications that deal with multiple frequency standards often use multiplexed clocks.

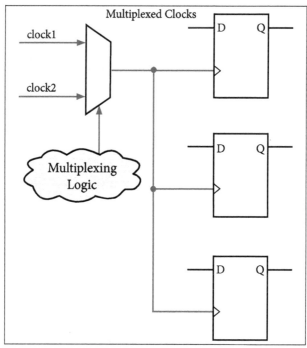

Multiplexed clock.

Clock multiplexing is acceptable if the following criteria are met:

- The clock multiplexing logic does not change after initial configuration.

- The design bypasses functional clock multiplexing logic to select a common clock for testing purposes.

- Registers are always in reset when the clock switches.

- A temporarily incorrect response following clock switching has no negative consequences.

- If the design switches clocks on the fly with no reset and the design cannot tolerate a temporarily incorrect response of the chip, then one must use a synchronous design so that there are no timing violations on the registers, no glitches on clock signals and no race conditions or other logical problems.

Synchronous Clock Enables and Gated Clocks

Gated clocks turn a clock signal on and off using an enable signal that controls some sort of gating circuitry. As shown in the below figure, when a clock is turned off, the corresponding clock domain is shut down and becomes functionally inactive.

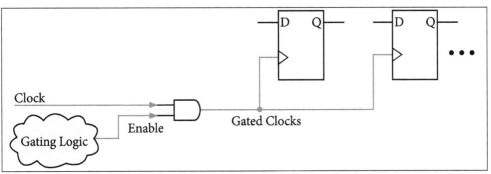

Gated clock.

Gated clocks can be a powerful technique to reduce power consumption. When a clock is gated, both the clock network and the registers driven by it stop toggling, thereby eliminating its contribution to power consumption.

However, gated clocks are not part of a synchronous scheme and therefore can significantly increase the effort required for design implementation and verification. Gated clocks contribute to clock skew and are also sensitive to glitches, which can cause design failure.

A clock domain can be turned off in a purely synchronous manner using a synchronous clock enable. However, when using a synchronous clock enable scheme, the clock tree keeps toggling and the internal circuitry of each flip flop remains active, which does not reduces the power consumption.

A synchronous clock enable technique is shown in the below figure. This Synchronous Clock Enable Clocking scheme does not reduces the power consumption as much as gating the clock at the source because the clock network keeps toggling, but it will perform the same function as a gated clock by disabling a set of flipflops.

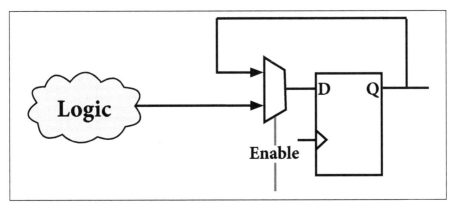

Synchronous clock enable.

As shown in the above figure, multiplexer in front of the data input of every flip flop either load new data or copy the output of the flipflop based on the Enable signal.

1.4.1 Classification of Memory Elements

Table: Semiconductor Memory Classification.

RWM		NVRWM	ROM
SRAM Access	Non - Random Access	EPROM	Mask - Programmed
SRAM	FIFO	E2PROM	Programmable(PROM)
DRAM	LIFO	FLASH	
	Shift Register		
	CAM		

Memory Arrays

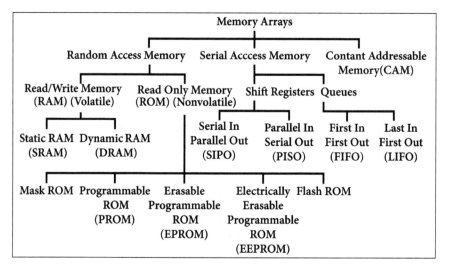

Memory Architecture: Decoders

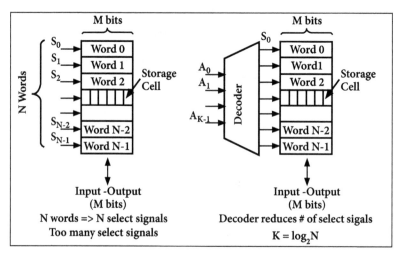

Array-Structured Memory Architecture

Array-Structured Memory Architecture has a problem that ASPECT RATIO or HEIGHT >> WIDTH.

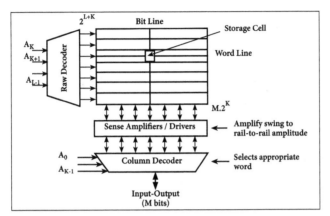

Hierarchical Memory Architecture

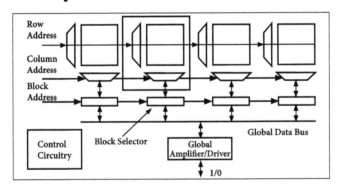

Advantages:

- Shorter wires within the blocks.

- Block address activates only 1 block and thus results in power saving.

Memory Timing: Definitions

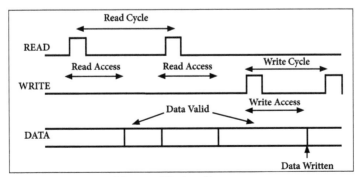

Memory Timing diagram.

Memory Timing: Approaches

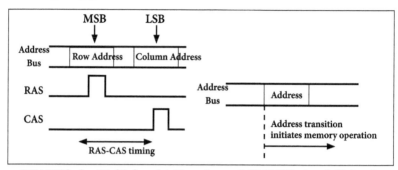

DRAM Timing Multiplexed Addressing and SRAM Timing Self-timed.

Read-Only Memories

- Read-Only Memories are non-volatile.

- Retain their contents when power is removed.

- Mask-programmed ROM's use one transistor per bit.

- Presence or absence determines 1 or 0.

Example:

4-word x 6-bit ROM

- Represented with dot diagram.

- Dots indicate 1's in ROM.

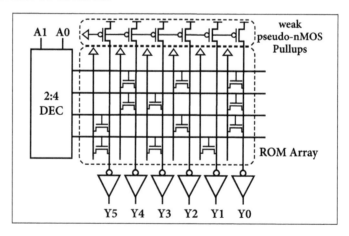

Word 0: 010101

Word 1: 011001

Word 2: 100101

Word 3: 101010

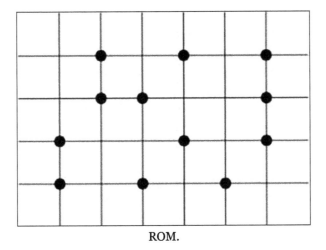

ROM.

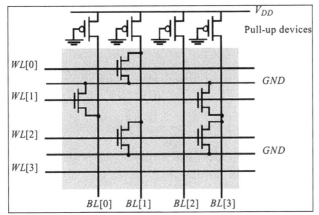

MOS NOR ROM.

Only 1 layer (contact mask) is used to program memory array. Programming of the memory can be delayed to one of the last process steps.

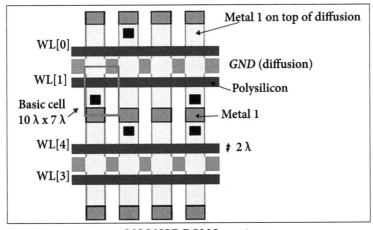

MOS NOR ROM Layout.

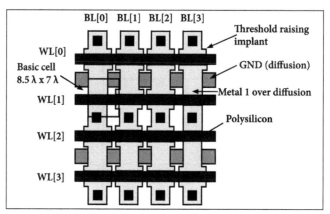

Threshold raising implants disable transistor.

Non-volatile Read-Write Memories (NVRW)

- Architecture is virtually identical to the ROM structure.

- The memory core consists of an array of transistors placed on a word-line (or) bit line grid.

- The memory is programmed by selectively disabling or enabling some of those devices.

- In a ROM, this is accomplished by mask level alterations.

- In a NVRW memory, the modified transistor permits its threshold to be altered electrically. The modified threshold is retained indefinitely even when the supply voltage is turned off.

- To reprogram the memory, the programmed values must be erased after which a new programming round can be started.

- The method of erasing is the main differentiating factor between the various classes of reprogrammable nonvolatile memories.

- Programming of memory is typically an order of magnitude slower than the reading operation.

PROM's and EPROM's

Programmable ROM's

- Build array with transistors at every site.

- Burn out fuses to disable the unwanted transistors.

Electrically Programmable ROM's

Uses floating gate to turn off the unwanted transistors.

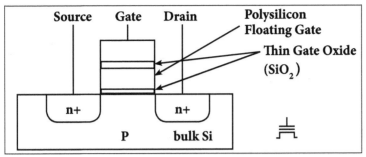

Transistor.

Floating-gate Transistor (FAMOS)

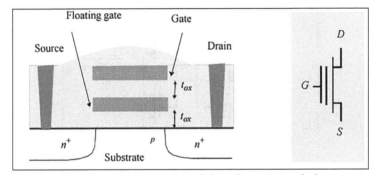

(a) Device cross-section and (b) Schematic symbol.

Characteristics of Non-Volatile Memory

	EPROM	EEPROM	Flash EEPROM
Memory size	16 Mbit(0.6μm)	1 Mbit (0.8μm)	16 Mbit(0.6μm)
Chip size	7.18 × 17.39 mm²	11.8 × 7.7 mm2	6.3 × 18.5 mm²
Cell size	3.8 μm2	30 μm2	3.4 μm²
Access time	62 nsec	120 nsec	58 nsec
Erasure time	minutes	N.A	4 sec
Programming time/word	5 μsec	8 msec/word, 4 sec/chip	5 μsec
Erase/Write cycles	100	105	10^3 - 10^5

Read-Write Memories

- STATIC RAM (SRAM)

- DYNAMIC RAM (DRAM)

Static Ram (Sram)

- Large (6 transistors/cell)

- Data stored as long as supply is applied

- Differential

- Fast

Dynamic Ram (Dram)

- Small (1-3 transistors/cell)
- Periodic refresh required
- Single Ended
- Slower
- 6-transistor CMOS SRAM Cell

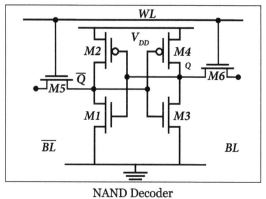

NAND Decoder

$$WL_0 = A_0A_1A_2A_3A_4A_5A_6A_7A_8A_9$$

$$WL_{511} = \overline{A}_0A_1A_2A_3A_4A_5A_6A_7A_8A_9$$

NOR Decoder

$$WL_0 = \overline{A_0 + A_1 + A_2 + A_3 + A_4 + A_5 + A_6 + A_7 + A_8 + A_9}$$

$$WL_{511} = \overline{\overline{A}_0 + \overline{A}_1 + \overline{A}_2 + \overline{A}_3 + \overline{A}_4 + \overline{A}_5 + \overline{A}_6 + \overline{A}_7 + \overline{A}_8 + \overline{A}_9}$$

Dynamic Decoders

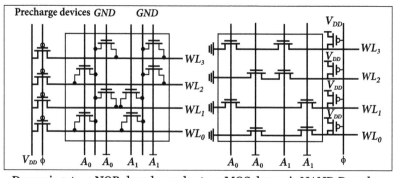

Dynamic 2-to-4 NOR decoder and 2-to-4 MOS dynamic NAND Decoder.

Programmable Logic Array

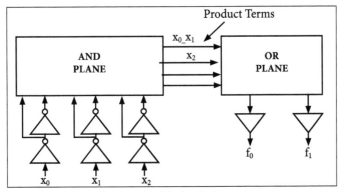

Programmable Logic Array.

A Programmable Logic Array performs any function in sum-of-products form.

- Literals: inputs & complements.

- Products/Minterms: AND of literals.

- Outputs: OR of Minterms.

Example: Full Adder

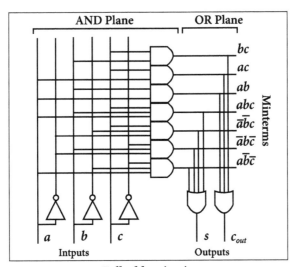

Full adder circuit.

$$s = a\overline{b}\overline{c} + \overline{a}b\overline{c} + \overline{a}\overline{b}c + abc$$

$$c_{out} = ab + bc + ac$$

Low Power Memory Circuits

The low-power RAM circuit is a major area of interest in low-power LSI research. Successive advances in the low-power RAM circuit have been able to suppress chip-power consumption, which increases with increasing memory capacity, chip area and speed. As a result, these advances coupled with high-density memory-cell technology have allowed chip power consumption to be maintained or lowered, although the memory capacity of DRAM chips has increased rapidly by over six orders (1 Kb to 4 Gb) over the past 30 years.

Consequently, the low-power RAM circuit has realized low-cost, high-reliability chips because it allows plastic packaging, a low operating current and a low junction temperature. In addition, it has managed the ever-increasing memory-subsystem power caused by the increasingly high throughput requirements of various processing systems such as personal computers.

Moreover, it forms not only the basis of other LSI memory chips, but also the basis of on-chip memory subsystems, such as embedded DRAMs (merged DRAM and logic) and SRAM caches, both of which have become increasingly important in modern memory systems.

RAM chip power is a prime concern of the subsystem designer, since it dominates memory-subsystem power. To reduce the active chip power, the designers of low-power RAM circuits have focused on three key issues so far such as reductions in the charging capacitance, operating voltage and dc current. Out of these issues, reduction in the operating voltage has become relatively important not only to reduce power, but also to ensure device reliability in scaled-down devices and also to extend the use of LSI's to battery-based portable systems.

Historically, DRAM designers initiated and then led the field in low-voltage LSI research, because their first priority was on higher-density chips, which were obtained through scaled-down FET's consequently resulting in lower breakdown voltages.

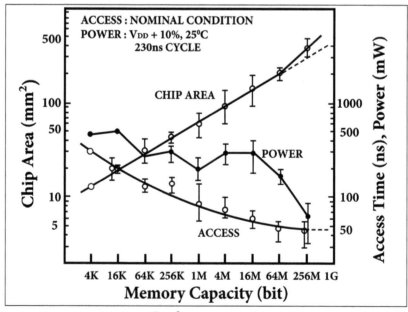

Graph area-memory.

Low-voltage (2-3V) circuits have been used in actual 16 Mb and 64 Mb DRAM products, although their external supply voltages are 5V or 3.3 V, being internally lowered by on-chip voltage down-converters to standardize the power supply.

Recent exploratory research on low-voltage operations of 3V or less suggests the great potential of CMOS circuits, although reducing voltage inevitably.

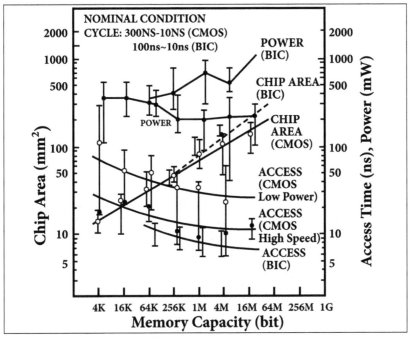

Graph area-memory.

The bit width for I/O pin is mostly 1 bit or 4 bits for DRAMs and high-speed SRAM's and mostly 8 bits for low-power SRAM's. DRAM and SRAM imposes memory-cell development, affording an active current of 23 mA at a 230 ns cycle time.

Reducing the data-retention power in DRAMs has been increasingly important for battery backup applications, since DRAMs are inherently less expensive to produce than SRAM's. A data-retention current as low as 3 AA at 2.6V has been reported for 4Mb chip as shown in the above figure. Such a low current finally allows the design of DRAM battery.

Basic Building Blocks, Current and Voltage Sources and CMOS Operational Amplifiers

2.1 Inverter with Active Load

Resistive-Load Inverters

The basic structure of a resistive load inverter is shown in the figure given below. Here, enhancement type NMOS acts as the driver transistor. The load consists of a simple linear resistor R_L. The power supply of the circuit is V_{DD} and the drain current I_D is equal to the load current I_R.

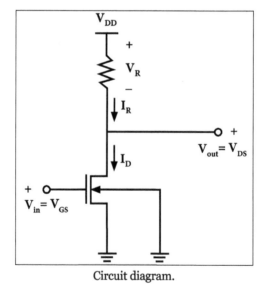

Circuit diagram.

Circuit Operation

When the input of the driver transistor is less than threshold voltage V_{TH} $(V_{in} < V_{TH})$, driver transistor is in the cut-off region and does not conduct any current. So, the voltage drop across the load resistor is ZERO and output voltage is equal to the V_{DD}. When the input voltage increases further, driver transistor will start conducting the non-zero current and NMOS goes in saturation region.

Mathematically,

$$I_D = \frac{K_n}{2}[V_{GS} - V_{TO}]^2$$

By increasing the input voltage further, driver transistor will enter into the linear region and output of the driver transistor decreases.

$$I_D = \frac{K_n}{2}2[V_{GS} - V_{TO}]V_{DS} - V_{DS}^2$$

VTC of the resistive load inverter, shown below. It indicates the operating mode of driver transistor and voltage points.

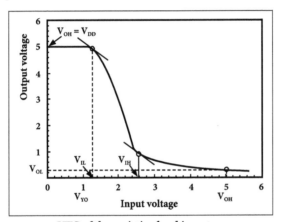

VTC of the resistive load inverter.

Inverters with n-Type MOSFET Load

In digital logic, an inverter or NOT gate is a logic gate which implements logical negation. The truth table is shown below:

V_{IN}	Q_2	$V_o = \overline{V_{IN}}$
0V (logic 0)	OFF	+5V (logic 1)
5V (logic 1)	ON	0V(logic 0)

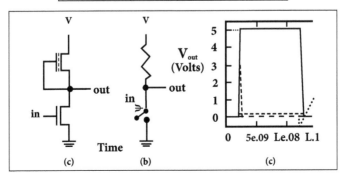

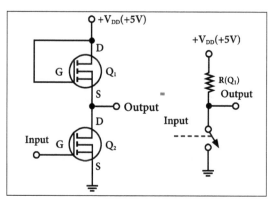

NMOS Inverter.

This represents perfect switching behavior, which is the defining assumption in Digital electronics. In practice, actual devices have electrical characteristics that must be carefully considered when designing inverters. In fact, the non-ideal transition region behavior of a CMOS inverter makes it useful in analog electronics as a class A amplifier.

The gate of the depletion mode transistor is connected to the drain, to keep the transistor permanently turned ON. The depletion mode transistor is used as a "pull-up" resistor and the enhancement mode transistor is used as a switch to "pull down" the output if the switch is turned ON.

It is important to note that in this technology, the resistance of permanently turned on depletion mode transistor should be great when compared with the "on" resistance of the enhancement mode transistor, but at the same time, small when compared with the "off" resistance of the transistor.

This type of logic is often known a "rationed logic", as the ratio of the pull-up resistance to the pull-down resistance effectively used to determine the voltage at which output of device changes state. Typically, $R_{pu} \approx 4R_{pd}$. The large resistive pull-up transistor results in three particular problems with this technology:

- The depletion mode transistor should be made large (i.e., long and thin) to create the large "on" resistance.

- When driving a capacitive output load like the gate of another transistor, the charging time (proportional to $R_{dep}C$) will be long when compared to the discharging time (which is proportional to $R_{enh}C$).

- The device eats up DC power whenever the enhancement mode pull down device is turned on, due to the resistive losses in pull-up transistor. The third problem becomes more serious. The third problem is that, as the feature sizes for transistors decreases, because the number of such resistors per unit area increases and the devices cannot dissipate the heat as well, resulting in the device failure as a result of overheating.

Estimation of Pull-up to Pull-down Ratio $\left(Z_{p.u}/Z_{p.d}\right)$ of an NMOS Inverter Driven by Another NMOS Inverter:

Figure below shows an inverter driven from the output of another similar inverter. Let VGS = 0 for the depletion mode transistor under all conditions. Also, in order to cascade inverters without having any inverse effect on levels, our target is to meet the requirement.

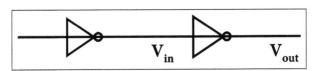

NMOS Inverter driven directly by another inverter.

$$V_{in} = V_{out} = V_{inv}$$

In order to have equal margins around the inverter threshold, we select $V_{inv} = 0.5V_{DD}$. Then both the transistors are in saturation. The drain-to-source current under saturation is given by,

$$I_{DS} = \frac{W}{L}K\frac{\left(V_{GS}-V_{t}\right)^{2}}{2}$$

Enhancement Mode

I_{DS} Pull-down device which is in the enhancement mode and has $V_{GS} = V_{inv}$ is given by,

$$I_{DS} = \frac{W_{p.d}}{L_{p.d}}K\frac{\left(V_{inv}-V_{t}\right)^{2}}{2}$$

Depletion Mode

The pull-up device which is in the depletion mode and has $V_{GS} = 0$ is given by,

$$I_{DS} = \frac{W_{p.u}}{L_{p.u}}K\frac{\left(V_{inv}-V_{t}\right)^{2}}{2} = \frac{W_{p.u}}{L_{p.u}}K\frac{\left(-V_{t}\right)^{2}}{2}$$

Since the two currents are same (p.u. and p.d. devices are in series), we have

$$\frac{W_{p.u}}{L_{p.u}}\left(-V_{td}\right)^{2} = \frac{W_{p.u}}{L_{p.u}}\left(-V_{inv}-V_{t}\right)^{2}$$

Where $W_{p.u}$, $L_{p.u}$, $W_{p.d}$ and $L_{p.d}$ are the widths and lengths of the pull-up and pull-down transistors respectively.

$$Z_{p.d} = \frac{L_{p.d}}{W_{p.d}} \text{ and } Z_{p.u} = \frac{L_{p.d}}{W_{p.d}}$$

We get, $\frac{1}{Z_{p.u}}(-V_{td})^2 = \frac{1}{Z}(V_{inv} - V_t)^2$

From which,

$$V_{inv} = V_t - \frac{V_{td}}{\sqrt{Z_{p.u}/Z_{p.d}}}$$

The typical values for the voltages are $V_{td} = -0.6\, V_{DD}$, $V_t = 01 V_{DD}$ and $V_{inv} = 0.5 V_{DD}$ (to have equal margins). Putting these values into equation we get,

$$0.5 = 0.2 + \frac{0.6}{\sqrt{Z_{p.u}/Z_{p.d}}}$$

From which,

$$\sqrt{\frac{Z_{p.u}}{Z_{p.d}}} = 2 \text{ or } \sqrt{\frac{Z_{p.u}}{Z_{p.d}}} = \frac{4}{1}$$

For an inverter driven directly by an inverter.

Alternative Forms of Pull-up

So far, we have seen two forms of pull-up where first in the form of a resistance and the second realized by a depletion mode pull-up device.

Two other possible forms of pull-up are as follows:

1. Complementary Transistor Pull-up (CMOS)

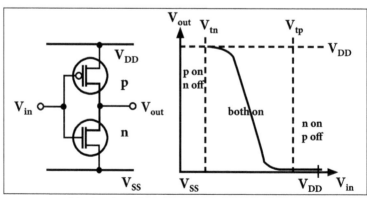

(a) Circuit and its (b) Transfer characteristics.
Complementary transistor pull-up (CMOS).

This form of pull-up is shown in the above figure. The output gives full logical 1 and 0 levels. No current flows either for logical 1 or for logical 0 inputs. For similar dimension devices, the n-channel device is faster than the p-channel device.

2. NMOS Enhancement Mode Pull-up

This configuration is shown in the figure below. In this circuit for output of logical 1, the V_{out} never reaches V_{DD}. Dissipation in the circuit is also high since current flows when V_{in} =logical 1. Dissipation can be considerably reduced if V_{GG} is obtained from a switching source such as from one phase of a clock. In this mode of pull-up, if V_{GG} is higher than VDD, an extra supply rail is required.

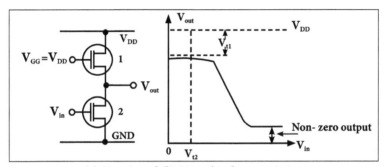

(a) Circuit and (b) Transfer characteristics.
Enhancement mode pull-up and transfer characteristics.

2.1.1 Cascode

The cascode is a two-stage amplifier composed of a trans-conductance amplifier followed by a current buffer. The word "cascode" was originated from the phrase "cascade to cathode". This circuit have a lot of advantages over the single stage amplifier like, better input output isolation, better gain, improved bandwidth, higher input impedance, higher output impedance, better stability, higher slew rate etc.

The reason behind the increase in bandwidth is the reduction of Miller effect. Cascode amplifier is generally constructed using FET (Field Effect Transistor) or BJT (Bipolar Junction Transistor). One stage will be usually wired in common source/common emitter mode and the other stage will be wired in common base/ common emitter mode.

Miller Effect

It is actually the multiplication of the drain to source stray capacitance by the voltage gain. The drain to source stray capacitance always reduces the bandwidth and when it gets multiplied by the voltage gain the situation is made further worse.

Multiplication of stray capacitance increases the effective input capacitance and as we know, for an amplifier, increase in input capacitance increases the lower cut of frequency and that means reduced bandwidth. It can be reduced by adding a current buffer stage at the output of the amplifier or by adding a voltage buffer stage before the input.

Compared to a single amplifier stage, this combination may have one or more of the following characteristics:

- High output impedance

- High input impedance

- High input-output isolation

- Higher gain or higher bandwidth

In modern circuits, cascode is often constructed from two transistors (BJT's or FET's) with one operating as a common emitter or common source and the other as a common base or a common gate. The cascode improves input-output isolation as there is no direct coupling from the output to input. This eliminates the Miller effect and thus contributes to a much higher bandwidth.

Operation

Figure shows an example of cascode amplifier with a common source amplifier as the input stage driven by signal source V_{in}. This input stage drives a common gate amplifier as output stage with output signal V_{out}.

As the lower FET is conducting by providing gate voltage, the upper FET conducts due to the potential difference now appearing between its gate and source.

The major advantage of this circuit arrangement stems from the placement of the upper field-effect transistor (FET) as the load of the input FET's output terminal. Because at operating frequencies the upper FET's gate is effectively grounded, the upper FET's source voltage is held at nearly constant voltage during operation.

In other words, the upper FET exhibits a low input resistance to the lower FET, making the voltage gain of the lower FET very small, which dramatically reduces Miller feedback capacitance from the lower FET's drain to gate. This loss of voltage gain is recovered by the upper FET.

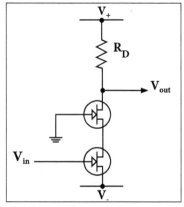

Cascode amplifier with a common source amplifier.

The upper transistor permits the lower FET to operate with minimum negative (Miller) feedback, improving its bandwidth. The upper FET gate is electrically grounded, so charge and discharge of stray capacitance C_{gd} between the drain and gate is simply through R_D and the output load and the frequency response is affected only for frequencies above the associated RC time constant:

$$\tau = C_{dg} R_D // R_{out}$$

Namely $f = 1/(2\pi\tau)$, a rather high frequency because C_{dg} is small. The upper FET gate does not suffer from the Miller amplification of C_{dg}.

If the upper FET stage were operated alone using its source as input node, it would have good voltage gain and wide bandwidth. However, its low input impedance would limit its usefulness to very low impedance voltage drivers. Adding the lower FET results in a high input impedance allowing cascode stage to be driven by a high impedance source.

If we have to replace the upper FET with a typical inductive/resistive load and take the output from the input transistor's drain, the CS configuration would offer the same input impedance as the cascode but the cascode configuration would offer a potentially greater gain and much greater bandwidth.

Practical Cascode Amplifier Circuit

A practical Cascode amplifier circuit based on FET is shown above. The Resistors R4 and R5 form a voltage divider biasing network for the FET Q2. R3 is the drain resistor for Q2 and it limits the drain current. The R2 is the source resistor of Q1 and C1 is its by-pass capacitor. R1 ensures zero voltage at the gate of Q1 during zero signal condition.

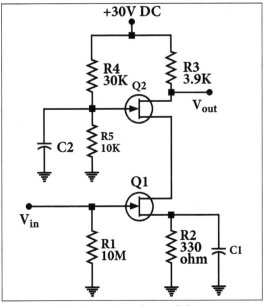

Practical Cascode amplifier.

2.1.2 Cascode with Cascode Load

The gain stages achieve their voltage gain because the signal current, $g_m v_{in}$ induced by the trans-conductance generator of the input transistor flows into a relatively high resistance (one r_{ds} or the parallel connection of two r_{ds}). This kind of operation is referred as trans-conductance gain amplification, because the active function of the gain component used is trans-conductor like.

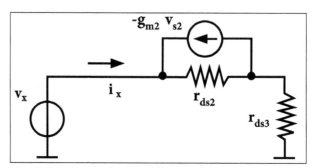

Equivalent circuit for the calculation of the impedance seen from the M2 source.

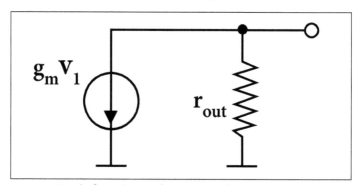

Equivalent circuit of a trans-conductance stage.

With the conventional design, the achieved gain $g_m r_{ds}$ ranges around 100. But for many practical applications, such a gain level cannot be sufficient. Therefore, to increase the gain while using a single stage amplifier, it would be necessary to enhance the trans-conductance or to augment the output resistance.

In saturation, an increase of the bias current or an enlargement of the aspect ratio magnifies the transistor's trans-conductance. However, augmenting the current depresses the output resistance by an amount that is larger than the benefit achieved. Therefore, the only viable way to raise the gain in a simple gain stage is to employ a very high aspect ratio in the input transistor. However, there are practical limits to this. Normally the maximum aspect ratio used is not extremely high.

A designer rarely uses several hundred for this ratio and one or two thousand are used only in very special cases. Hence, a limited increase is expected in the gain by merely pushing up the aspect ratio. In contrast, increasing the output resistance can lead to better results.

The resistance seen from the output drain of a cascode is larger than the output resistance of a simple transistor by a pretty large factor, $g_m r_{ds}$. This result suggests that we should use a cascode arrangement even on the active load side, as shown in the below figure.

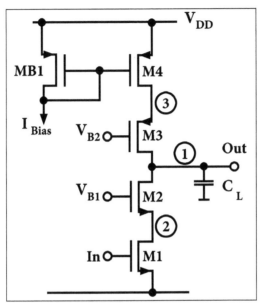

Cascode with cascode load.

The resistance of the output node becomes much higher, as it is the parallel-connection of two cascode structures. Since the output voltage is given by the product of the signal current and the output resistance, a rough expression of the dc voltage gain becomes,

$$A_v = -g_{m1} \frac{\left(r_{ds1} g_{m2} r_{ds2}\right)\left(r_{ds4} g_{m3} r_{ds3}\right)}{r_{ds1} g_{m2} r_{ds2} + r_{ds4} g_{m3} r_{ds3}}$$

The dc gain of a cascode with cascode load is approximately the square of the gain of an inverter with active load.

It is worth noting that the gain is proportional to the square of the product of a trans-conductance and a transistor output resistance. So, the gain is almost the square of a simple inverter with active load. If the transistors are in saturation, the gain of the simple inverter is inversely proportional to the square root of the bias current. Therefore, for the cascode with cascode load, the gain is inversely proportional to the bias current.

As for the inverter, the gain increases when the current decreases down until the point at which the transistors enters the sub-threshold region, the trans-conductance becomes proportional to the bias current and the dc gain reaches its maximum, typically 80 to 100 dB.

An important result if the cascode with cascode load has a simple configuration, it achieves the same gain as the cascade of two inverters with active load. This advantage is counterbalanced by a disadvantage that a reduced output swing. Here, we have to design two bias voltages, one for the gates of M2 and the other for the gate of M3. To choose these, we have to remember that the values of V_{B1} and V_{B2} should allow a suitable margin to keep transistors M1 and M4 in saturation, but at the same time, they should minimize the reduction of the output swing.

The maximum achievable dynamic range in a cascode with cascode load corresponds to $2V_{sat}$ distance from the supply voltages. Moreover, the bias voltages used must track the threshold fluctuations.

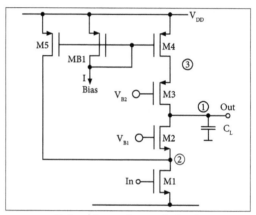

Gain enhanced cascade load.

In the figure, the additional current is injected into the input transistor to improve the dc gain.

Therefore, for optimum output dynamic range, we should use the following design guidelines for V_{B1} and V_{B2}.

$$V_{B1} \sim V_{sat,1} + V_{GS2} + \Delta = V_{Th,n} + 2V_{sat} + \Delta$$

$$V_{B2} \sim V_{DD} - V_{sat,4} - V_{GS3} - \Delta = V_{DD} - V_{Th,p} - 2V_{sat} - \Delta$$

Where Δ is the suitable margin necessary for accommodating the possible mismatch of the threshold voltages and taking the variations of the bias current into account.

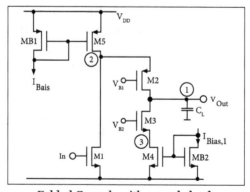

Folded Cascode with cascode load.

The cascode with cascode load adequately achieves high dc gain. Nevertheless, the circuit can be improved upon and the schematic can be rearranged to better meet specific needs. The two modified versions of the cascode with cascode load are the gain - enhanced version and the folded version.

2.2   Source Follower

Constant Current Source

The common source amplifier is the most widely used because it provides a high-impedance input and both voltage gain and current gain. The source follower in contrast, provides no voltage gain. The output follows the input a VT drop below. It does, however provide a high-impedance input and it is very fast. Finally, the cascode amplifier provides no current gain.

When the input is below $V_{GG} - VT_n$ the input and output are effectively connected and the input receives the entire output load. The cascode does however, provide voltage gain.

Source Follower

A source follower is most often used with a current-source load. It is used as a level shifter or as a buffer for a small-swing signal driving a large capacitive load. In this configuration, the output signal follows the input signal with a drop given by,

$$V_{out} = V_{in} - V_{Tn} - \left(\frac{l_s}{\beta}\right)^{1/2}$$

Because all current for falling edges is provided by the current source, the speed of the source follower is limited to,

$$t_d = \frac{\Delta V C_L}{l_S}$$

Where ΔV is the voltage swing and C_L is the capacitive load on the output. Thus as I_s is made larger, the source follower gets faster but the voltage drop for a give n transis.

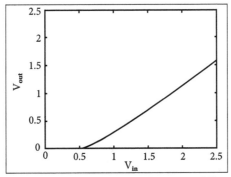

Simulated DC characteristics of a source follower.

Because the FET is in saturation, almost all of the gate capacitance is to the source, which is moving in the same direction as the input. Thus, the charge across C_G remains nearly constant across the operating range of the circuit and the effective input capacitance is very small. This is, In essence, this is the flip-side of the Miller effect.

When the gain is positive and nearly unity, shows that the effective feedback capacitance vanishes results in no gain, no capacitive pain. Because of this effect, the FET in a source follower can be made quite wide without excessively loading its input. Logic of a sort can be built out of source followers. This circuit computes the function $D = (A \wedge B) \vee C$ with three caveats.

First, the circuit is not composable. Because output D is a voltage drop, VGT and has a slightly smaller swing than the input signals, it must be amplified and level-shifted before it can be used as an input to a similar circuit.

Second, the reduced input capacitance of the source follower is seen only by the input that causes the output to switch. For example if A, B and C are all high and B swings low, it sees the full CG of its FET because the output D remains roughly constant.

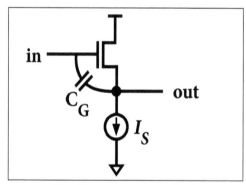

(a) with current-source load.

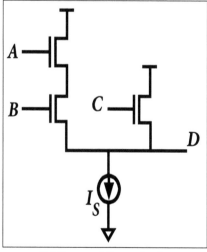

(b) Source-follower logic.

There are three basic configurations of IC MOSFET amplifiers:

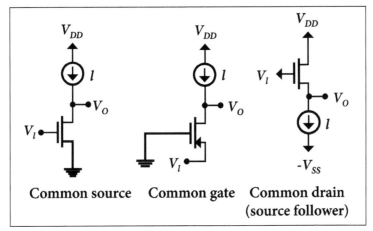

Common source.

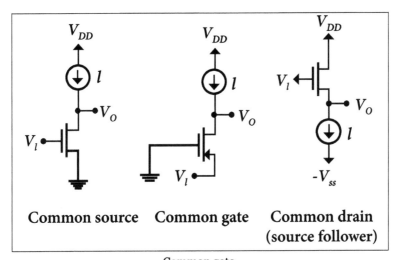

Common gate.

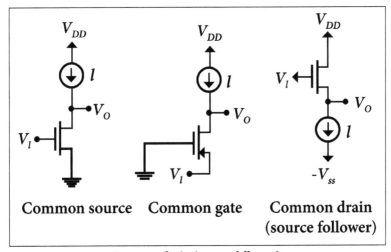

Common drain (source follower).

Large-valued resistors and capacitors are not often used in these IC environments. Instead active loads are incorporated using MOSFETs as loads. In the amplifier circuits shown above, the active loads are the non-ideal current sources. Similar to the emitter follower, the source follower can be analyzed as a resistor divider.

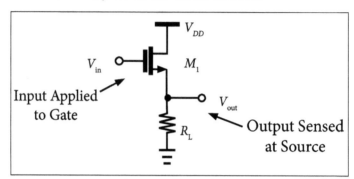

Source Follower Stage

2.2.1 Threshold Independent Level Shift

The threshold dependency and the relatively high shift achieved are not suitable for many applications. With a typical technology used, the source follower shift is in the order of 1V and because of the technological variations; the inaccuracy can be as high as some hundreds of mV. In many cases, the designer needs a small voltage shift, threshold independent. So, the source follower cannot be used.

Complementary configurations achieve the same function. The circuit in the below figure (a) has a diode connected transistor (M1) at the input which shifts the input voltage upwards than a source follower M2 shifts the result downwards. Instead, the circuit in figure (b) has a source follower M2 before and the diode connected transistor (M1) after.

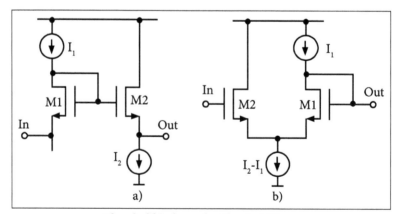

Threshold independent level-shifters.

The small signal behavior results from a direct inspection of the circuits. We can verify that the small signal resistance between input and output is approximated by,

$$r_{eq} = 1/g_{m1} + 1/g_{m2} \qquad \qquad \qquad ...(1)$$

Therefore for the small signals, the above circuit's acts like the source follower with an additional output series resistance equal to $1/g_{m1}$.

Both the solutions achieve the voltage shift through the series of a shift-up and a shift down. By inspection of the two circuits, we have,

$$\Delta V = V_{Th,I} + \sqrt{\frac{2I_{DI}}{\mu C_{ox} \left(\dfrac{W}{L}\right)_I}} - V_{Th,2} - \sqrt{\frac{2I_{D2}}{\mu C_{ox} \left(\dfrac{W}{L}\right)_2}} \qquad ...(2)$$

Since the two shifts are due to the same kind transistors, the thresholds will match and the result is just the difference between two overdrives.

$$\Delta V = \frac{2}{\mu C_{ox}} \cdot \left\{ \sqrt{I_{DI} / \left(\frac{W}{L}\right)_I} - \sqrt{I_{D2} / \left(\frac{W}{L}\right)_2} \right\} \qquad ...(3)$$

In addition to a threshold independency, the circuits achieve the level shift value by the proper choice of transistor aspect ratios and bias currents. Therefore, designer can achieve positive or negative and even pretty small voltage shifts.

2.2.2 Improved Output Stages

We have seen that the simple source follower provides output resistances in the $k\Omega$ range. This is normally enough for the charge or the discharge of low-medium capacitive loads. However, the performances achieved become poor when the buffer is required to drive the resistive loads or large capacitors at high frequency.

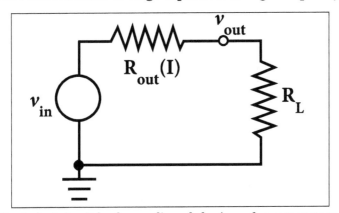

Equivalent circuit for the non-linear behaviour of an output stage.

For this design situation, the output stage should provide the output resistance significantly lower than the bad impedance that it drives. Moreover, to keep the harmonic distortion under control, the variation of output resistance induced by current swings must be a very small fraction of the load.

For better understanding, let us consider the equivalent circuit. It represents the equivalent circuit of an output stage driving a resistive load. The output voltage is given by,

$$V_{out} = V_{in} \frac{R_L}{R_{out} + R_L} \qquad \qquad ...(4)$$

Assume that $R_{out} = R_{out,o}(1 + \alpha(1))$ and a α suitable function expressing the non-linear behavior can be represented as,

$$V_{out} = V_{in} \frac{R_L}{R_{out,o} + R_L} \left\{ 1 - \frac{R_{out,o}\alpha(I)}{R_{out,o} + R_L} \right\} \qquad ...(5)$$

The output voltage is, therefore, an attenuated replica of the input. But in addition, we have the term $R_{out,o}\alpha(1)/R_{out,o} + R_L$ which is responsible for a non-linear response. Many applications can tolerate the attenuation of the signal. Instead, non-linearity's are not admissible beyond a given limit since they produce harmonic distortion.

In order to prevent this sort of problem, the designer can follow two possible strategies to desensitize the non-linearity effect, by decreasing $R_{out,o}$ so that it becomes much smaller than R_L or to improve the output resistance linearity.

2.3   Current and Voltage Sources: Current Mirrors and References

Current mirror

The circuit in which the output current is forced to equal the input current is called as current mirror circuit. In a current mirror circuit, the output current is the mirror image of input current. The basic block diagram is shown in figure (a), while figure (b) shows the circuit diagram.

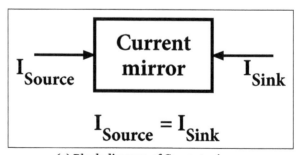

(a) Block diagram of Current mirror.

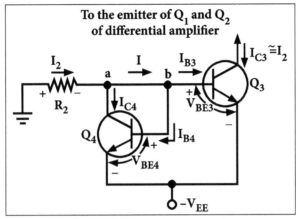

(b) Circuit diagram of Current mirror.

Circuit Analysis:

The circuit consists of two matched transistors Q_3 and Q_4. Their base emitter voltage and base currents are same.

$$\therefore V_{BE3} = V_{BE4} \text{ and } I_{B3} = I_{B4}$$

Similarly, their collector currents are also same.

$$\therefore I_{C3} = I_{C4}$$

Applying KCL at node a,

$$I_2 = I_{C4} + 1 \qquad\qquad ...(1)$$

Applying KCL at node b,

$$I = I_{B3} + I_{B4} = 2I_{B4} = 2I_{B3} \qquad\qquad ...(2)$$

$$\therefore I_2 = I_{C4} + 2I_{B4} = I_{C3} + 2I_{B3} \qquad\qquad ...(3)$$

Now, $IB_3 = \dfrac{I_{C3}}{\beta}$

$$\therefore I_2 = I_{C3} + I_{C3}\left(\dfrac{2}{\beta}\right) \qquad\qquad ...(4)$$

Generally β is very large and hence $(2/\beta)$ is negligibly small.

$$\therefore I_2 = I_{C3} \qquad\qquad ...(5)$$

Thus the collector current Q_3 of is nearly equal to the current I_2 Hence once current mirror circuit is set for current, I_2 it provides the constant current bias to the differential amplifier. Thus, I_2 can be obtained by writing KVL for the base-emitter loop of transistor Q_3.

$$-I_2 R_2 - V_{BE3} + V_{EE} = 0$$

$$\therefore I_2 = \frac{V_{EE} - V_{BE3}}{R_2}$$

Selecting R_2, the appropriate I_2 can be set for the current mirror circuit.

Improved Current Mirror Circuit

The modified current mirror circuit is shown in the figure below, I_2 cannot be equal to IC3 when β is small.

Circuit Analysis:

Applying KCL at node 'b' we get,

$$I_1 = I_{C1} + I_{B3} \qquad \qquad ...(7)$$

The two transistors Q_1 and Q_2 are identical.

$$I_{B1} = I_{B2} = I_B \qquad \qquad ...(8)$$

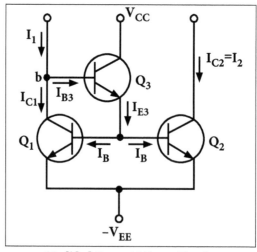

Modified Current Mirror Circuit.

Hence, the emitter current I_{E3} of transistor Q_3 gets divided equally.

$$\therefore I_{E3} = 2I_b \qquad \qquad ...(9)$$

Now, $I_{E3} = (1 + \beta) I_{B3}$

$$\therefore 2I_B = (1 + \beta)I_{B3} \qquad \qquad ...(10)$$

Now, $I_{C1} = I_{E1} = \beta I_B$ ...(11)

And $I_{C2} = I_{E2} = \beta I_B$ ...(12)

Substituting (11) and (10) in (7), we get,

$$I_1 = \beta I_B + \frac{2I_B}{(1+\beta)}$$

$$I_1 = I_B \left[\frac{\beta(1+\beta)+2}{(1+\beta)} \right] \qquad \qquad ...(13)$$

Substituting in (12) we get,

$$I_{C2} = \beta \left[\frac{I_1(1+\beta)}{\beta(1+\beta)+2} \right] \qquad \qquad ...(14)$$

Thus,

$$I_{C2} = I_{C1} = I_2 = \frac{I_1\beta(1+\beta)}{\beta(1+\beta)+2} \qquad \qquad ...(15)$$

The ratio I_2/I_1 is much less dependent upon β for this modified circuit.

Wilson Current Source Circuit

Wilson current source provides an output current Io which is equal to I_{ref}. High output resistance is the feature of Wilson current source. The Wilson current source circuit is shown in the figure below.

The high output resistance is achieved due to the negative feedback through Q_3. It also provides base currents cancellation making Io nearly equal to I_{ref}.

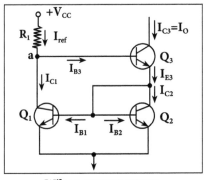

Wilson current source.

Circuit Analysis:

Applying KCL at node 'a'

$$I_{ref} = I_{C1} + I_{B3} \qquad \qquad ...(16)$$

And $I_{E3} = I_{C2} + I_{BI} + I_{B2}$ \qquad \qquad ...(17)

But as all the transistors Q_1, Q_2 and Q_3 are identical, their base-emitter voltages and base currents are also equal.

$$\therefore I_{B1} = I_{B2} = I_{B3} = I_B \qquad \qquad ...(18)$$

$$\therefore I_{E3} = I_{C2} + 2I_{B2}$$

Also $I_{E3} = I_{C3} + I_{B3}$ ...(19) \qquad \qquad ...(19)

i.e., $I_{E3} = I_{C3} + I_B$

Now, $I_{C3} = I_{E3} - I_B$

$$= I_{c2} + 2I_b - I_b$$

$$= I_{C2} + I_b \qquad \qquad ...(20)$$

But, $I_{C2} = I_{C1}$ as transistors are identical.

$$\therefore I_{C3} = I_{C1} + I_B \qquad \qquad ...(21)$$

Comparing the equations (21) and (16), we can say that,

$$I_o = I_{C3} = I_{ref}$$

Now, $I_{E3} = I_{C2} + 2I_B$

But, $I_B = \dfrac{I_{C2}}{\beta}$

$$\therefore I_{E3} = I_{C2} + \dfrac{2I_{C2}}{\beta}$$

$$I_{E3} = I_{C2}\left[1 + \dfrac{2}{\beta}\right] \qquad \qquad ...(22)$$

Now, $I_{C3} = I_{E3}\left[\dfrac{\beta}{1+\beta}\right]$ ...(23)

Substituting (22) in (23), we get

$$I_{C3} = I_{C2}\left[1+\dfrac{2}{\beta}\right]\left[\dfrac{\beta}{1+\beta}\right]$$ ...(24)

$$I_{C2} = I_{C3}\dfrac{1}{\left[1+\dfrac{2}{\beta}\right]\left[\dfrac{\beta}{1+\beta}\right]}$$ ...(25)

But as two transistors are identical, $I_{C2} = I_{C1}$

From equation (16), we can write

$$I_{ref} = I_{C2} + I_{B3}$$

i.e. $I_{C1} = I_{ref} - I$

$$I_{C1} = I_{ref} - \dfrac{I_{C3}}{\beta}$$ ...(26)

So from the equations (25) and (26),

$$I_{C3}\dfrac{1}{\left[1+\dfrac{2}{\beta}\right]\left[\dfrac{\beta}{1+\beta}\right]} = I_{ref} - \dfrac{I_{C3}}{\beta}$$

$$\therefore\ I_{C3}\left[\dfrac{(1+\beta)\beta}{\beta(2+\beta)}+\dfrac{1}{\beta}\right] = I_{ref}$$

$$\therefore\ I_{C3}\left[\dfrac{\beta^2+\beta+2+\beta}{\beta(2+\beta)}\right] = I_{ref}$$

$$\therefore\ I_{C3}\left[\dfrac{\beta^2+2\beta+2}{\beta(2+\beta)}\right] = I_{ref}$$

$$\therefore\ I_{C3} = I_{ref}\left[\dfrac{\beta(2+\beta)}{\beta^2+\beta+2}\right]$$

$$\therefore\ I_{C3} = I_{ref}\left[\dfrac{\beta^2+2\beta+2-2}{\beta^2+2\beta+2}\right]$$

$$I_{C3} = I_{ref} \left[1 - \frac{2}{\beta^2 + 2\beta + 2} \right]$$

...(27)

The equation shows that the output current $(I_{C3} = I_o)$ and the reference current I_{ref} differ only by a factor which is of the order of $2\beta^2$. Hence, the low sensitivity of the circuit towards the transistor base currents is verified. The output resistance of the Wilson current mirror is greater than the simple current mirror.

Widlar Current Source

In the operational amplifier, low input current is required. Hence input stage is biased at very low current, typically at a collector current of the order of 5 µA. Currents of such low magnitude can be obtained with a modified circuit called Widlar current source. The circuit is shown in the figure below.

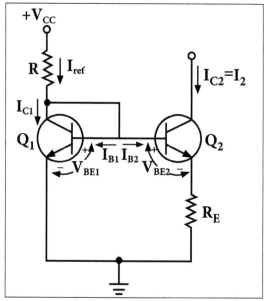

Widlar current source.

Circuit Analysis:

The two transistors are identical to each other, but due to the emitter resistance R_E, V_{BE1} and V_{BE2} are different. In fact, $V_{BE2} < V_{BE1}$ and hence $I_{C2} < I_{C1}$. Due to the asymmetric nature of the base emitter loop, the circuit is called 'lens' rather than a 'mirror'.

Applying KVL to the base emitter loop,

$$V_{BE1} = V_{BE2} + \left(I_{B2} + I_{C2} \right) R_E$$

...(28)

i.e., $V_{BE1} = V_{BE2} = \left(I_{B2} + I_{C2} \right) R_E$

$$V_{BE1} = V_{BE2} = \left(I + \frac{I}{\beta}\right) I_{C2} R_E \qquad \qquad ...(29)$$

For a transistor, we can write,

$$I_{C1} = I_S \, e^{V_{BEI}/V_T}$$

And

$$I_{C2} = I_S \, e^{V_{BE2}/V_T}$$

Where I_s is the reverse saturation current, same for the two transistors and V_T is voltage equivalent of temperature. These equations are called as Embers-Moll equations for the transistor.

$$\frac{I_{C1}}{I_{C2}} = e^{(V_{BEI} - V_{BE2})/V_T} \qquad \qquad ...(30)$$

Taking natural logarithm of both the sides, we get

$$In\left(\frac{I_{C1}}{I_{C2}}\right) = \frac{V_{BEI} - V_{BE2}}{V_T}$$

Hence, we get

$$V_{BEI} - V_{BE2} = V_T \, In\left(\frac{I_{C1}}{I_{C2}}\right) \qquad \qquad ...(31)$$

Equating equations (29) and (31),

$$(I + \beta) IC_2 R_E = V_T In\left(\frac{I_{C1}}{I_{C2}}\right) \qquad \qquad ...(32)$$

$$R_E = \frac{V_T}{\left(1 + \frac{I}{\beta}\right) IC_2} In \frac{IC_1}{IC_2} \qquad \qquad ...(33)$$

Writing KCL of the collector point of Q_1,

$$I_{ref} = I_{C1} + I_{B1} + I_{B2}$$

$$IC_1 = \left(1 + \frac{I}{\beta}\right) + \frac{IC_2}{\beta}$$

In the Widlar current source, $I_{C_2} \ll I_{C_1}$. Therefore, the term I_{c2}/β may be neglected.

Thus, $I_{ref} = IC_1\left(1 + \frac{I}{\beta}\right)$

$$IC_1 = \frac{\beta}{\beta+1}I_{ref}, \text{ where } I_{ref} = \frac{V_{CC} - V_{EE}}{R_I}$$

For design purposes, I_{C_1} and I_{C_2} are usually known and equation (6) provides the value of R_E required to achieve the desired value of I_C.

Use of an Active Load to Improve CMRR

To improve CMRR, it is necessary to increase A_d.

While, $A_d = \dfrac{h_{fc}R_C}{R_S + h_{ie}}$

Thus to increase A_d, R_C must be high as possible.

But there are limitations to select maximum value of R_C such as:

- For large R_C, the quiescent drop is more hence higher biasing voltage is necessary to maintain the quiescent collector current.

- Higher value of R_C, requires a large chip area. Hence it is not possible to increase the value of R_C beyond a particular limit.

The current mirror circuit has very low d.c. resistance $\left(\dfrac{dV}{dI}\right)$ and higher a.c. resistance $\left(\dfrac{dv}{di}\right)$.

The requirement to increase the gain is same that the collector resistance should not disturb d.c. conditions while it must provide large resistance for a.c. purposes. Hence, the current mirror circuit can be used as a collector load instead of R_C. Such a load is called an active load.

The quiescent voltage across the current mirror is the fraction of the supply voltage. This eliminates the need of high biasing supply voltage. It basically acts as a current source and provides large a.c. resistance.

The differential amplifier using a current mirror as an active load is shown in the below figure.

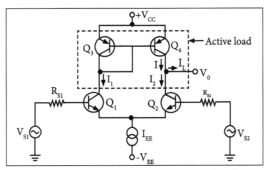

Differential amplifier with an active load.

Under the d.c. conditions, $V_{S1} = V_{S2} = 0$. Q_1 As and Q_2 are matched transistors hence, $I_1 = I_2 = I_{EE/2}$ where base currents of Q_1 and Q_2 are neglected. The transistors Q_3 and Q_4 form a current repeater hence $I = I_1 = 1_2$.

The load current IL entering the next stage is,

$$I_L = I - I_2 = 0$$

But when V_{S1} increases over V_{S2}, the current I_1 increases whereas I_2 decreases as $I_1 + I_2 = I_{EE}$ constant. Also, the current I always remain equal to I_1 due to the current mirror action. Thus, the active load provides very high a.c. resistance and hence high differential mode voltage gain. Thus, as Ad becomes high, CMRR gets improved.

2.3.1 Voltage Biasing

The simplest way to generate a bias voltage is a diode-connected transistor depicted in figure (a). It is generally used for a simple current mirror biasing or a cascode transistor biasing. The generated voltage VBIAS depends on the bias current IBIAS and is given by,

$$V_{BIAS} = V_{To} + 2nV_t \cdot \ln\left[\exp\left(\sqrt{\frac{I_{RIAS}}{I_S}}\right) - I\right] \qquad ...(1)$$

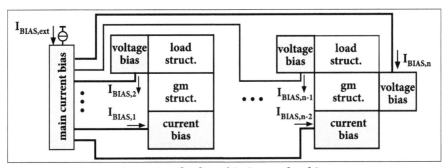

Current and voltage biasing on the chip.

If a simple current mirror is biased, the diode-connected transistor sizes are matched with the active load transistors.

$$\frac{(W/L)_{\text{diode}}}{(W/L)_{\text{mirror}}} = \frac{I_{\text{BIAS}}}{I_{\text{mirror}}} \qquad \qquad \dots(2)$$

Where the transistor length is same for all the transistors and the ratio of the transistor widths corresponds to the ratio of the currents.

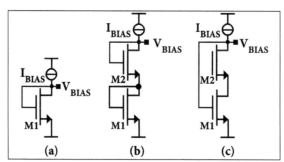

Topologies for voltage bias generation: a) Diode-connected transistor, b) Two stacked diode-connected transistors, c) Diode-connected transistor with the equivalent resistance in the source.

If a large DC bias voltage has to be generated, two diode-connected transistors can be stacked as shown in figure (b), Now $V_{\text{BIAS}} > 2V_{\text{To}}$. However, the most often used topology shown in figure (c), makes it possible to have $V_{\text{To}} < V_{\text{BIAS}} < 2V_{\text{To}}$. Here, the transistor M1 operates in the linear region and acts as an equivalent resistance in the source of diode-connected transistor M2.

Finally, when a small bias voltage is needed, the topology depicted in the below figure can be used. The bias voltage is then calculated as follows. The transistors M1 and M2 connected in series act as an equivalent transistor whose sizes are calculated as,

$$\frac{L_{\text{eq}}}{W_{\text{eq}}} = \frac{L_1}{W_1} + \frac{L_2}{W_2} \qquad \qquad \dots(3)$$

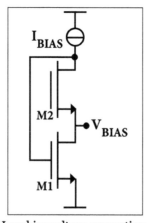

Low bias voltage generation.

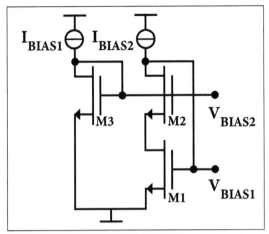

Cascode current mirror bias voltage generation.

Thus,

$$V_{BIAS} = 2nV_t \cdot \ln\left\{\left[\exp\left(\sqrt{\frac{I_{BIAS}}{I_{S,eq}}}\right)-1\right]\Big/\left[\exp\left(\sqrt{\frac{I_{BIAS}}{I_{S2}}}\right)-I\right]\right\} \qquad ...(4)$$

Where,

$$I_{S,eq} = 2nKP\left(\frac{W_{eq}}{L_{eq}}\right)V_t^2 \text{ and } I_{S2} = 2nKP\left(\frac{W_2}{L_2}\right)V_t^2$$

The above figure displays a voltage bias suitable for the cascode current mirror. Transistors M1 and M2 are matched with the cascode current mirror transistors, i.e., the transistor lengths are the same and the transistor widths are calculated from the ratio of the currents. The transistor M3 is sized to generate V_{BIAS2} that is required to keep the transistor M1 in the saturation region.

Voltage Source

A voltage source is a circuit that produces an output voltage V_o, which is independent of the load driven by the voltage source or the output current supplied to the load.

The voltage source is the circuit dual of the constant current source. A number of IC applications require a voltage reference point with very low AC impedance and stable DC voltage which is not affected by the power supply and temperature variations. There are two methods which can be used to produce the voltage source such as:

- Using the impedance transforming properties of the transistor, which determines the current gain of the transistor.

- Using an amplifier with negative feedback.

Voltage Source Circuit using Impedance Transformation

The voltage source circuit using the impedance transforming property of the transistor is shown in the figure below. The source voltage V_s drives the base of the transistor through the series resistance R_s and the output is taken across the emitter.

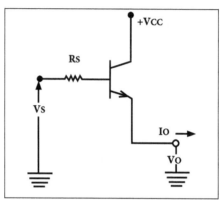

Voltage source circuit using Impedance transformation.

The load regulation parameter indicates the changes in V_o resulting from large changes in output current I_o. Reduction in V_o occurs as I_s goes from no-load current to full-load current and this factor determines the output impedance of the voltage sources.

Common Collector Type Voltage Source

The voltage follower (or) common collector type voltage source is suitable for the differential gain stage used in op-amps. Figure below shows an emitter follower or common-collector type voltage source.

The low output impedance of the common-collector stage simulates a low impedance voltage source with an output voltage level of Vorepresented by,

$$V_o = \left(\frac{R_2}{R_1 + R_2} \right) V_{cc}$$

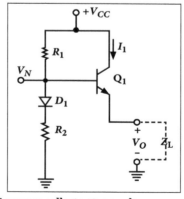

Common collector type voltage source.

The diode D, is used for offsetting the effect of DC value VBE across the emitter-base junction of transistor and for compensating the temperature of V_{BE} drop of Q_1.

Advantages:

- Producing low ac impedance.

- Resulting in effective decoupling of adjacent gain stages.

Voltage Source using Temperature Compensated Avalanche Diode

The limitation of voltage source using emitter follower is that the output voltage V_o changes with respect to changes in supply voltage V_{CC}. This can be overcome in the voltage source circuit using the breakdown voltage of the base emitter junction shown in the figure below:

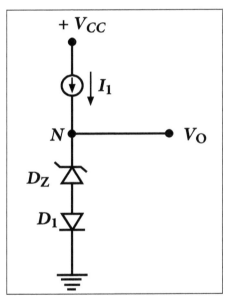

Voltage source using the breakdown voltage of the base-emitter junction.

The circuit eliminates the need of emitter follower because the impedance looking into the bias terminal N is very low. The current source I_1 is simulated by a resistor connected between V_{CC} and node N. Then the output voltage level V_o at node N is given by,

$$V_o = V_z + V_{BE}$$

Where,

V_B - The breakdown voltage of diode D_Z.

V_{BE} - The voltage drop across the diode D_Z.

The diode D_1 provides partial temperature coefficient effect of V_Z.

In a monolithic IC structure, D_2 and D_1 can be realized as a single transistor with two emitter as shown in the figure below.

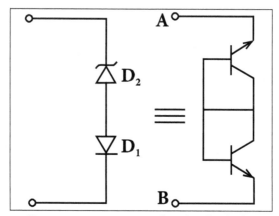

Temperature compensated avalanche diode.

Since the transistors have their base and collector terminals common, they can be designated as a single transistor with two emitters.

2.3.2 Voltage References

The circuit that is primarily designed for providing a constant voltage independent of changes in temperature is termed as voltage reference. The most important characteristic of a voltage reference is the temperature coefficient of the output reference voltage TCR.

The desirable properties of a voltage reference are:

- Reference voltage must have good power supply rejection which is as independent of the supply voltage as possible.

- Reference voltage must be independent of any temperature change.

- The voltage reference circuit is used to bias the voltage source circuit and the combination can be called as the voltage regulator. The basic design strategy is producing a zero TCR at a given temperature and thereby achieving good thermal ability. Temperature stability of the order of 1000ppm°C is typically expected.

- Output voltage must be as independent of the loading of the output current as possible or in other words, the circuit should have low output impedance.

Voltage Reference Circuit using Temperature Compensation Scheme

Voltage reference circuit using the basic temperature compensation scheme is shown below. This design utilizes the close thermal coupling achievable among the monolithic components and this technique compensates the known thermal drifts by introducing an opposing and compensating drift source of equal magnitude.

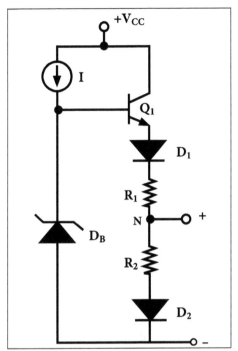

Voltage Reference circuit using temperature compensation.

2.4   CMOS Operational Amplifiers: General Issues

- Infinite voltage gain

- Infinite input impedance

- Infinite bandwidth

- Zero input offset voltage (i.e., exactly zero out if zero in)

- Zero output impedance

AMP Characteristics

The characteristics of an op-amp are:

Supply Voltage

Two types of supply are used for op-amps, the dual and single supply. Many op-amps, especially older types use dual supply $(+V_S$ and $-V_S)$ often in the 12 to 18V range.

Single voltage supplies have grown in popularity with the increase in portable devices where dual supplies using multiple batteries are more expensive to implement.

Frequency Response

In the ideal amplifier, its frequency response should be infinite; it will amplify all the frequencies equally. In practical amplifiers, this is difficult to achieve and not always desirable but the op-amps have extremely wide and easily variable bandwidths.

Open Loop Voltage Gain

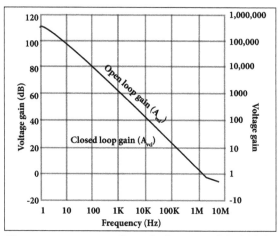

Frequency response of a typical op-amp (LMC660).

The above graph shows that the open loop gain with no feedback at very low frequencies is huge. Open loop gain refers to the maximum AC gain at very low frequencies. The graph shows that open loop voltage gain of about 126dB, but at frequencies above a few Hz, gain begins to fall rapidly at 20dB/decade.

Large Signal Voltage Gain

This is the open loop voltage gain measured at DC with the amplifier producing the large voltage output.

Closed Loop Voltage Gain

Practically, huge gain of an op-amp is greatly reduced by applying an appropriate amount of the negative feedback. Dotted line shows the response of the op-amp with negative feedback. The gain has been reduced to 20dB, a closed loop voltage gain $\left(A_{vcl}\right)$ of x10, which has produced a flat response from 0Hz to 140kHz.

Gain Bandwidth Product

The term 'Gain Bandwidth Product' is often used to describe the possible combinations of gain and bandwidth. Bandwidth indicated by the Gain Bandwidth Product applies to small signals.

For large AC signals, with fast rising and falling edges, the bandwidth may be further reduced by the Slew Rate. Then, Power Bandwidth becomes more relevant.

Maximum Differential Input

This is the maximum voltage that can be applied between the two inputs. In some devices, this can be equal to the supply voltage.

Input Resistance

It is the resistance looking into input terminals with the amplifier operating without feedback (open loop). For bipolar devices, it ranges from 1MΩ to 10MΩ. For FET and CMOS types, it ranges up to 1012Ω or more.

Input Offset Current

The currents flowing into the two inputs should be ideally zero. But for practical op-amps, small input currents exist and may also be different. Unequal currents cause different voltages at inputs.

When this small difference in voltage is amplified, it causes the output to be other than zero. To overcome this effect, an Input Offset Voltage can be applied between the inputs to correct the output voltage to zero. Typical values for bipolar op-amps would be ±1mV ranging up to 15mV for FET types.

Temperature Coefficients

The input offset current and input offset voltage are affected by the changes in temperature. The temperature coefficient of input offset current is measured in nA or pA/°C. The temperature coefficient of input offset voltage is usually measured in µV/°C .

Slew Rate

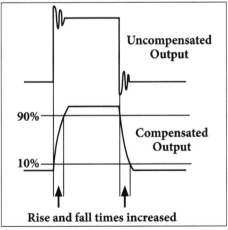

Slew Rate.

The slew rate describes how fast the output voltage can change in response to an immediate change in voltage at the input. Higher the value (in V/µs) of slew rate, faster the output can change.

If a square wave is applied to the input, the output should also be a square wave but the fast rising and falling edges of the square wave causes the amplifier to oscillate for a short time after the rise or fall. To prevent this, the op-amp's internal circuitry contains a small amount of compensation capacitance.

It slows down the rate of change by acting as a time constant. This compensation also limits slew rate of the op-amp. If the slew rate of the amplifier can't keep up with fastest rate of change of the signal, some distortion will be produced.

For amplifying the large amplitude signals, an op-amp needs to have sufficiently high value of slew rate to cope with greatest possible rate of voltage change. If the largest possible voltage swing and the highest frequency of the signal are known, the minimum required slew rate for the op-amp can be calculated using the formula,

Slew rate $(V / \mu s) = 2\pi f V_{pk}$

Where,

> f = The highest signal frequency (Hz).

> V_{pk} = The maximum peak voltage of the signal.

Power Bandwidth

In an op-amp operating from a ±15V supply, power bandwidth would be specified as the frequency range in which a ±10V swing can be measured at output with a total harmonic distortion of less than 5%.

Ideal Operational Amplifier

It has two input terminals and one output terminal. Other terminals have not been shown for simplicity. The - and + symbols at the input refer to inverting and non-inverting input terminals respectively, i.e., $v_1 = 0$ output v_o is 180° out of phase with input signal v_2 and when $v_2 = 0$, output v_o will be in phase with the input signal applied at v_1.

This op-amp is said to be ideal if it has the following characteristics:

Open loop voltage gain, $A_{OL} = \infty$

Input impedance, $R_i = \infty$

Output impedance, $R_o = 0$

Bandwidth, $BW = \infty$

Zero offset, i.e. $v_o = 0$ when $v_1 = v_2 = 0$

It can be seen that:

An ideal op-amp draws no current at both the input terminals i.e., $i_1 = i_2 = 0$. Because of infinite input impedance, any signal source can drive it and there is no loading on the preceding driver stage.

Since gain is ∞, the voltage between the inverting and non-inverting terminals, i.e., differential input voltage $v_d = (v_1 - v_2)$ is essentially zero for finite output voltage v_o.

The output voltage v_o is independent of the current drawn from the output as $R_o = 0$. Thus, the output can drive an infinite number of other devices.

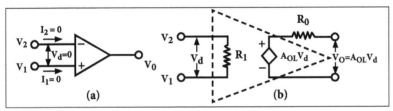

(a) Ideal op-amp, (b) Equivalent circuit of an op-amp.

The above properties can never be realized in practice. However, the use of such an 'Ideal op-amp' model simplifies the mathematics involved in op-amp circuits. There are practical op-amps that can be made to approximate some of these characteristics.

A physical amplifier is not an ideal one. So, the equivalent circuit of an op-amp may be shown in Figure (b) where $A_{OL} \neq \infty$, $R_i \neq \infty$, $R_o \neq 0$. It can be seen that op-amp is a voltage controlled voltage source and A_{OL}, v_d is an equivalent Thevenin voltage source and Ro is the Thevenin equivalent resistance looking back into the output terminal of an op-amp.

The equivalent circuit is useful in analyzing the basic operating principles of op-amp. For the circuit shown in Figure (b), the output voltage is given by,

$$v_o = A_{OL} v_d = A_{OL} (v_1 - v_2)$$

The equation shows that the op-amp amplifies the difference between the two input voltages.

2.4.1 Performance Characteristics

An ideal op-amp draws no current from source and its response is also independent of temperature. But in a real op-amp, current is taken from the source to input terminals. The two inputs respond differently to current and voltage due to mismatch in transistors. The non-ideal characteristics are:

- Input bias current
- Input offset current

- Thermal drift

- Output offset voltage

Input Bias Current

The Op-amp's input is a differential amplifier, which may be made of BJT or FET. In either case, the input transistor must be biased into their linear region by supplying currents into the bases by the external circuit.

In ideal Op-amp, no current is drawn from input terminals. But practically, op-amp input terminals do conduct a small value of AC current to bias input transistors. Manufacturers specify the input bias current IB as the average value of base currents entering into the terminals of an op-amp as in the below figure.

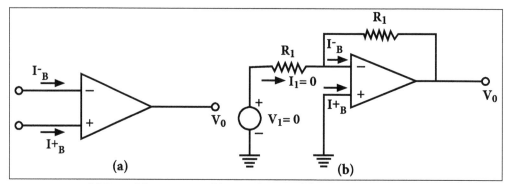

(a) Input bias currents, (b) Inverting amplifier with bias currents.

Bias current is the input current. The transistor (internal transistor of the input differential stage) needs to produce an amount of output voltage.

$$I_B = \frac{I_B^+ + I_B^-}{2} \qquad \qquad ...(1)$$

For 741, a bipolar op-amp, the bias current is 500 nA or less. Consider the basic inverting amplifier. If the input voltage V_i is set to zero volts, V_O should be zero volts.

Instead, we find that the output voltage is offset by,

$$V_O = (I_B) R_f \qquad \qquad ...(2)$$

For Example:

If R_f = 1MΩ, I_B = 500 nA, then V_O = 500 mV with zero input. Output voltage is 500 mV. This is unacceptable. This effect can be compensated as in figure by adding a resistor R_{comp} between the non-inverting input terminal and ground.

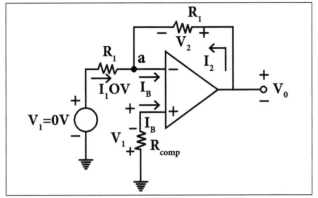

Bias current compensation in an inverting amplifier.

Current I_B^+ flowing through R_{comp} develops a voltage V_1 across it.

Then by KVL, we get,

$$-V_1 + 0 + V_2 - V_0 = 0$$

$$V_0 = V_2 - V_1 \qquad \qquad ...(3)$$

By selecting proper value of R_{comp} V_2 can be canceled with V_1 and output V_O will be zero.

$\therefore$ The value of R_{comp} is derived as,

$$V_1 = I_B + R_{comp}$$

$$I_B^+ = \frac{V_1}{R_{comp}} \qquad \qquad ...(4)$$

The node 'a' is at voltage $\left(-V_1\right)$, because the voltage at the non-inverting input terminal is $\left(-V_1\right)$. So with $V_i = 0V$, we get,

$$I_1 = \frac{V_i - \left(-V_1\right)}{R_1} = 0 + \frac{V_1}{R_1} = \frac{V_1}{R_1} \qquad \left[\because V_i = 0\right]$$

$$I_1 \frac{V_1}{R_1} \qquad \qquad ...(5)$$

Also,

$$I_2 \frac{V_2}{R_f} \qquad \qquad ...(6)$$

For compensation, V_O should be zero for $V_i = 0V$ i.e., $V_1 = V_2$

$$\therefore I_2 = \frac{V_1}{R_f} \qquad \qquad ...(7)$$

KCL at node 'a' gives,

$$\Rightarrow I_1 + I_2 = I_B^-$$

Substituting for I_1 and I_2,

$$I_B^- = \frac{V_1}{R_1} + \frac{V_1}{R_f}$$

$$\therefore I_B^- = V_1 \frac{(R_1 + R_f)}{R_1 R_f} \qquad \qquad ...(8)$$

Assuming $I_B^- = I_B^+$ and using equation (4) and (8) we get,

$$I_B^+ = \frac{V_1}{R_{copm}}$$

$$\therefore V_1 \frac{(R_1 + R_f)}{R_1 R_f} = \frac{V_1}{R_{comp}} \qquad \qquad ...(9)$$

$$R_{comp} = R_1 \parallel R_f = \frac{R_1 R_f}{R_1 + R_f}$$

To compensate bias currents, R_{comp} should be equal to parallel combination of resistors R_1 and R_f. The effect of input bias current in an amplifier can be compensated by placing the compensating resister R_{comp}.

Input Offset Current

The difference in magnitudes of I_B^+ and I_B^- is called as input offset current (I_{os}). Bias current compensation will work if both I_B^+ and I_B^- are equal. Since the input transistors are not identical, there is a small difference between I_B^+ and I_B^-. This difference is called as input offset current (I_{os}) and can be written as,

$$|I_{os}| = I_B^+ - I_B^- \qquad \qquad ...(10)$$

The absolute value of sign indicates that there is no way to predict which of the bias currents will be larger.

Offset current I_{os} for BJT op-amp is 200nA. Offset current I_{os} for FET is 10pA. Even with bias current compensation, I_{os} will produce a output voltage when $V_i = 0V$.

$$V_1 = I_B^+ R_{comp} \qquad \qquad ...(11)$$

$$\text{And } I_1 = \frac{V_1}{R_1} = \frac{I_B^+ R_{comp}}{R_1} \qquad \qquad ...(12)$$

KCL at node 'a' $\Rightarrow I_1 + I_2 = I_B^-$

$$I_2 = I_B^- - I_1$$

$$I_2 = \left(I_B^- - I_1\right) = I_B^- - \left(I_B^+ \frac{R_{comp}}{R_1}\right) \qquad \qquad ...(13)$$

$$V_o = V_2 - V_1$$

$$V_o = I_2 R_f - V_1$$

$$V_o = I_2 R_f - I_B^+ R_{comp}$$

$$= \left(I_B^- - I_B^+ \frac{R_{comp}}{R_1}\right) R_t - I_B^+ R_{comp} \qquad \qquad ...(14)$$

Substitute for $Rc_{omp} = R_1 || R_f$

$$V_o = I_B^- R_f - I_B^+ R_{comp}\left[\frac{R_f}{R_t} + 1\right]$$

$$= I_B^+ R_f - I_B^+ \frac{R_t R_f}{R_1 + R_f}\left[\frac{R_1 + R_f}{R_1}\right]$$

$$V_o = R_f\left[IB^- - IB^+\right] \qquad \qquad ...(15)$$

$$V_o = R_f I_{os} \qquad \qquad ...(16)$$

So even with current bias compensation and with feedback resistor of 1MΩ, a 741 BJT Op-amp has an output offset voltage,

$$V_O = 1M\Omega \times 200nA$$

$$= 200mV, \text{ with zero input voltage.}$$

It can be seen that effect of offset current can be minimized by having a small value of R_f. Unfortunately; R_1 must be kept large to have high input impedance. With R_1 large, R_f also should be kept large so as to have reasonable gain.

This will allow large feedback resistance, while keeping the resistance to ground (seen by the inverting input) low as shown in dotted in the network.

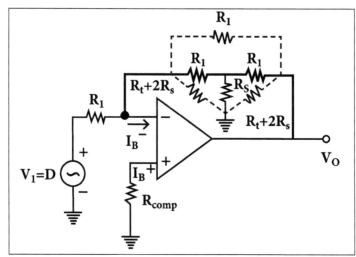

Inverting amplifier with 'T' feedback network.

The 'T' Network provides a feedback signal as if the network were a single feedback resistor.

By T to π conversion,

$$R_f = \frac{R_t^2 + 2R_t R_s}{R_s} \qquad \qquad \dots(17)$$

$$R_f R_s = R_t^2 + 2R_t R_s$$

$$R_t^2 = R_f R_s - 2R_t R_s$$

$$R_t^2 = R_s \left(R_f - 2R_t \right)$$

To design a T-network,

$$R_t \ll R_f / 2 \qquad \qquad \dots(18)$$

Then calculate $R_s = \dfrac{R_t^2}{R_f - 2R_t}$ $\qquad \qquad \dots(19)$

Input Offset Voltage

Whenever both the input terminals of the op-amp are ground ideally, the output voltage should be zero. However, the practical op-amp shows a small non-zero output voltage.

To make this output voltage zero, one may have to apply a small voltage at the input terminals. This voltage is called input offset voltage V_{os}, since positive terminal (+) voltage is around $V_1 = 0V$.

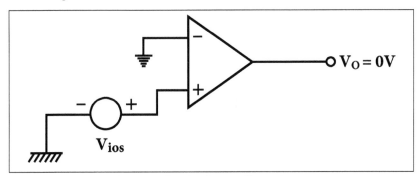

Op-amp showing input offset voltage.

The voltage V_2 at (-) input terminal is given by,

$$V_2 = \left(\frac{R_1}{R_1 + R_f}\right) V_0 \qquad \qquad ...(20)$$

$$V_0 = \left(1 + \frac{R_1 + R_f}{R_1}\right) V_2$$

$$V_0 = \left(1 + \frac{R_f}{R_1}\right) V_2 \qquad \qquad ...(21)$$

Since $V_{iOS} = |V_1 - V_2| \qquad \qquad ...(22)$

And $V_1 = 0V$

$$VOS = |0 - V_2| = V_2$$

$$\therefore \ V_0 = \left(1 + \frac{R_f}{R_1}\right) V_{iOS} \qquad \qquad ...(23)$$

Thus, the output offset voltage of an op-amp in closed loop configuration (inverting and non-inverting) is given by equation (23).

Total Output Offset Voltage

The total output offset voltage VoT could be either more or less than the offset voltage, produced at the output due to input bias current or input offset voltage alone. This is because the input offset voltage V_{ios} and input bias current I_B could be either positive or negative with respect to ground.

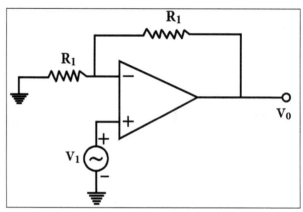

(a) Non Inverting amp.

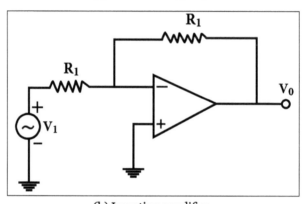

(b) Inverting amplifier.

Without any compensation technique used, the maximum offset voltage at output of inverting and non-inverting amplifier of figure (a) and (b) is,

$$V_{oT} = \left(1 + \frac{R_f}{R_1}\right) V_{ios} + R_f I_B \qquad \qquad ...(24)$$

With R_{comp} in the circuit, the total output offset voltage will be given by,

$$V_{oT} = \left(1 + \frac{R_f}{R_1}\right) V_{OS} + R_f . I_{ios} \qquad \qquad ...(25)$$

Many op-amps provide offset compensation pins to nullify the offset voltage. Figure gives the connections for 741 op-amp.

The manufacturers recommend a 10 kΩ potentiometer be placed across offset null pins. 1 and 5 and the wiper is connected to the negative supply pin 4. The position of wiper is adjusted to nullify the output offset voltage.

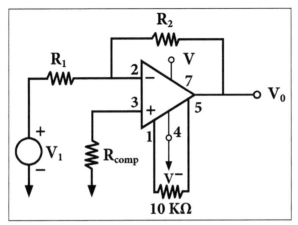

Offset null pin connection for μA741.

For small signal sinusoidal (AC) applications, one has to know the AC characteristics such as frequency response and slew rate.

Frequency Response of an Op-Amp

Ideally, an open loop op-amp has infinite bandwidth. It means that, if its $A_{OL} = 90dB$ with DC signal, its gain should remain the same 90dB through audio and onto high frequencies. But with practical op-amp, gain decreases (rolls off) at higher frequencies.

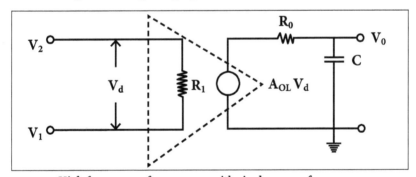

High frequency of an op-amp with single corner frequency.

The reason for the above is that there must be a capacitance component in the equivalent circuit of op-amp. This capacitance is due to the physical characteristics of device (BJT or FET) used and internal construction of op-amp. For the op-amp with one corner frequency, all the capacitor effects can be represented by a single capacitor C as in diagram.

Due to single RoC, there is one pole and obviously one -20dB/decade rolls off comes into effect. The open loop gain of an op-amp with only one corner frequency is obtained from the figure as,

$$v_o = \frac{-jXc}{R_o - jXc} A_{OL} \cdot Vd \qquad \qquad ...(26)$$

$$A = \frac{V_o}{V_d} = \frac{A_{OL} \cdot 1}{1 + j2\pi f R_o C} \qquad \ldots (27)$$

$$A = \frac{A_{OL}}{1 + j(f/f_1)}$$

Or, $\quad A = \dfrac{A_{OL}}{\sqrt{1 + j\left(\dfrac{f}{f_1}\right)^2}}$

Where, $\quad A = \dfrac{A_{OL}}{\sqrt{1 + j\left(\dfrac{f}{f_1}\right)^2}} \qquad \ldots (28)$

Where f1 - corner frequency of the op-amp.

The magnitude and phase angle of the open loop voltage gain can be written as,

$$|A| = \frac{A_{OL}}{\sqrt{1 + \left(\dfrac{f}{f_1}\right)^2}} \qquad \ldots (29)$$

$$\phi = -\tan^{-1}\left(\frac{f}{f_1}\right) \qquad \ldots (30)$$

The magnitude and phase characteristics are as shown in figure.

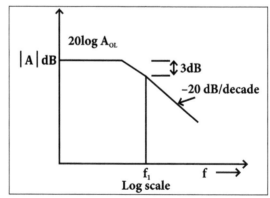

Open loop magnitude characteristics in semi-log paper.

From the graph,

- For frequency $f < f_1$, the magnitude of the gain is $20 \log A_{OL}$ in dB.

- For $f = f_1$, the gain is 3 dB down from the dc value of A_{OL} in dB. This frequency f_1 is called corner frequency.

- For $f \gg f_1$, the gain rolls off at the rate of -20 dB/decade (or) 6dB/octave.

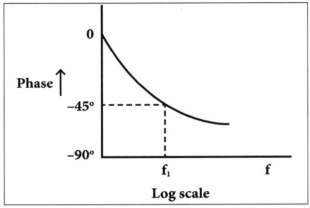

Phase Characteristics for an op-amp with single break frequency.

From the phase Characteristics, the phase angle is zero at frequency $f = 0$. At f_1, phase angle is -45° (lagging) and at infinite frequency phase angle is -90°. This shows that a maximum of 90° phase change can occur in an op-amp with a single capacitor.

The voltage transfer function in s-domain can be written as,

$$A = \frac{A_{OL}}{1 + j\dfrac{f}{f_1}}$$

$$= \frac{A_{OL}}{1 + j(\omega / \omega_1)} = \frac{A_{OL} \cdot \omega_1}{j\omega + \omega_1}$$

$$= \frac{A_{OL} \cdot \omega_1}{S + \omega_1}$$

Due to the number of RC pole pairs, there will be a number of different break frequencies. The transfer function of an op-amp with 3 different break frequencies can be assumed as,

$$A = \frac{A_{OL}}{\left(1 + j\dfrac{f}{f_1}\right)\left(1 + j\dfrac{f}{f_2}\right)\left(1 + j\dfrac{f}{f_3}\right)} \qquad \text{...(31)}$$

$$0 \geq f_1 < f_2 < f_3$$

$$A = \frac{A_{OL} \cdot \omega_1 \omega_2 \omega_3}{(s + \omega_1)(s + \omega_2)(s + \omega_3)} \qquad \text{...(32)}$$

With $0 < \omega_1 < \omega_2 < \omega_3$

Approximation of open loop gain V_s frequency curve is shown in the figure below:

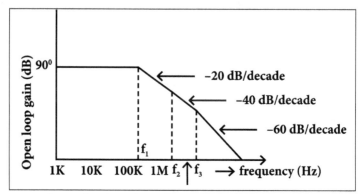

Approximation of open loop gain Vs frequency curve.

Approximation of open loop gain Vs frequency curve is shown in the figure. For frequencies from 0 Hz to 200 KHz, open loop frequency response is constant (90 dB).

From 200 kHz - 2 MHz, gain drops from 90 dB to -70 dB which is at a -20 dB/decade. From 2 MHz - 20 MHz - the roll off rate is -40 dB/decade. As frequencies increases, cascading effect of RC pairs (poles) come into effect and roll-off rate increases successively by -20 dB/decade at each corner frequency.

Slew Rate

The maximum rate of change of output voltage with respect to time is called slew rate of the Op-amp.

$$S = \frac{dV_o}{dt}\bigg|_{max}$$

$$S = 2\pi f\, V_m \; V/sec$$

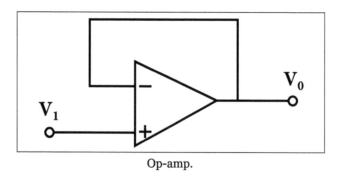

Op-amp.

The slew rate limits the speed of response of all the large signal wave shapes, V_i as a large amplitude, high frequency sine-wave with a peak amplitude of V_m.

$$V_o = V_m \sin \omega t$$

$$V_o = V_m \sin \omega t$$

The rate of change of output is given by,

$$\frac{dV_o}{dt} = V_m \omega \cos \omega t$$

The maximum rate of change of the output occurs when cost $\omega t = 1$.

Slew rate $\left. \dfrac{dV_o}{dt} \right|_{max} = V_m \omega$

The maximum frequency f_{max}, at which an undistorted output voltage with a peak value V_m can be obtained is determined by,

$$f_{max} = \frac{Slew\,rate}{2\pi V_m}$$

The maximum peak sinusoidal output voltage V_m that can be obtained at a frequency of f is given by,

$$V_{m(max)} = \frac{Slew\,rate}{2\pi f}$$

Methods of improving slew rate:

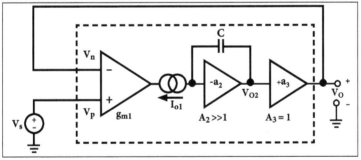

Op-amp model for slew rate analysis.

Gain bandwidth product is given by,

$$f_1 = \frac{gm_1}{2\pi C}$$

It is also called as unity gain bandwidth of an op-amp,

$$S = 2\pi \frac{I_{o1}(sat)}{g_{m1}} f_1$$

Method 1: Increasing f_{ts}

To increase f_t, internal capacitor value must be reduced. Compensating network area is used to reduce 'C'.

Method 2: Increasing I_{o1} (sat)

I_{o1} (sat) is increased by using the programmable op-amps, where the operating point of the device can be sent by the user with external current I_{set}.

Method 3: Reducing g_{m1}

- By using suitable resistance in series with the emitter of the different input transistors.

- By using FET differential input pair instead of BJT.

2.5   Basic Architecture

Differential Gain Stages of CMOS Op-Amp Architecture

Differential amplifiers are typically used as the input stages of op-amps. The purpose of such differential stages is to amplify the difference of two signals. Moreover, they should provide large power-supply rejection ratio (PSRR), large common-mode rejection ratio (CMRR), high input impedance, low noise and offset voltage since all the parasitic effects such as offset or noise are amplified in the following amplifying stages.

A typical CMOS differential amplifier which can fulfil the above requirements is shown in figure (a). It consists of two source-coupled p-channel MOS transistors M_1, M_2 biased by a current source and an NMOS current mirror (transistors M_3, M_4) used as an active load.

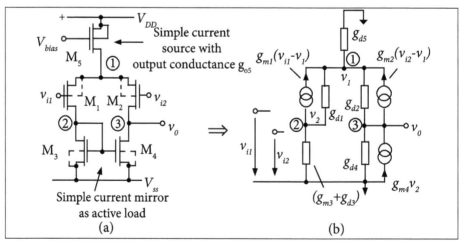

(a) CMOS differential gain stage using p-channel input devices and
(b) its small-signal low-frequency equivalent circuit.

For proper operation, the transistors M_1, M_2 and M_3, M_4 must be precisely matched. The low-frequency small-signal equivalent circuit of the differential stage is shown in figure (b). It is assumed that the current source has the internal output resistance $r_{o5} = 1/g_{o5}$.

The circuit shown in figure (b) was obtained by the inspection of the differential stage of figure (a) considering the equivalent models of the MOS transistors. Note that the body effect of MOSFETs does not occur since, the substrates are connected to the corresponding sources of the MOS transistors, as indicated by dotted lines in figure (a). An approximate analysis of the circuit can be readily performed by applying KCL to the nodes 1, 2 and 3:

$$g_{m1}\left(v_{i1}-v_1\right)+g_{d1}\left(v_2-v_1\right)+g_{m2}\left(v_{i2}-v_1\right)+g_{d2}\left(v_o-v_1\right)-g_{o5}v_1 = 0$$

$$-g_{m1}\left(v_{i1}-v_1\right)+g_{d1}\left(v_1-v_2\right)+\left(g_{m3}-g_{d3}\right)v_2 = 0$$

$$g_{m4}v_2+g_{d4}v_o+g_{d2}\left(v_o-v_1\right)+g_{m2}\left(v_{i2}-v_1\right)=0 \qquad \text{...(1)}$$

Taking into account that the transistor pairs are ideally matched

(i.e., $g_{m1}=g_{m2}$, $g_{d1}=g_{d2}$, $g_{m3}=g_{m4}$ and $g_{d3}=g_{d4}$) and in practice $g_{mi} \gg g_{di}$ for i = 1,2,3,4, we can find the approximate solution of the system of equations (1) in the form,

$$v_o \cong \frac{g_{m1}}{g_{d1}+g_{d3}}\left(v_{i1}-v_{i2}\right)=\frac{g_{o5}g_{d1}}{2g_{m3}\left(g_{d1}+g_{d3}\right)}\left[\frac{v_{i1}-v_{i2}}{2}\right] \qquad \text{...(2)}$$

Hence, we can write,

$$A_o = A_d := \frac{v_o}{v_{i1}-v_{i2}}\bigg|_{v_{i1}+v_{i2}=0} \cong \frac{g_{m1}}{g_{d1}+g_{d3}} \gg 1 \qquad \text{...(3)}$$

$$A_{CM} := \frac{v_o}{\left(v_{i1}-v_{i2}\right)/2}\bigg|_{v_{i1}+v_{i2}=0} \cong \frac{g_{o5}g_{d1}}{2g_{m3}\left(g_{d1}+g_{d3}\right)} \ll 1 \qquad \text{...(4)}$$

$$CMRR := \frac{|A_d|}{|A_{CM}|} \cong \frac{2g_{m1}g_{m3}}{g_{o5}g_{d1}} \gg 1 \qquad \text{...(5)}$$

From the above analysis, it follows that to achieve high CMRR, the current source must have a large output resistance $r_{o5} = 1/g_{o5}$ (i.e., a "good" current source is required). Moreover, the trans-conductances of the transistors M1 ÷ M4 must be sufficiently large to provide a high voltage (differential) gain.

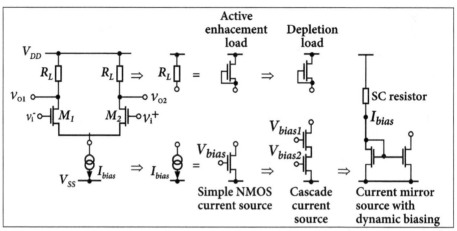

CMOS differential amplifiers using n-channel input devices: (a) Double-ended differential amplifiers.

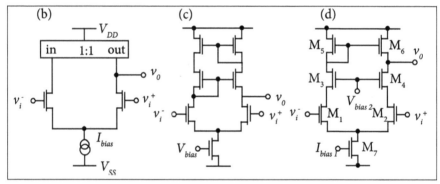

Single-ended differential amplifiers: (b) General configuration, (c) Differential stage with a Wilson current mirror as load, (d) Differential stage cascode load.

Figure (a) shows a conventional double ended (i.e., two-output) differential stage. A single MOSFET operating in the saturation region can be used as the biasing current source. By connecting two MOSFETs in series, we can increase the small-signal resistance of the current source. The increase of this resistance results in a more constant current source for a variation of the output voltage across it.

Another method to implement the current source in MOS technology is to employ a current mirror which can be biased through a resistor realized in a switched capacitor (SC) technique. This kind of biasing is called dynamic biasing and is used when very low dc power dissipation is imperative. Such biasing is especially suited for SC applications.

A simple way to realize the resistive load in MOS technology is to apply the enhanced-mode MOSFETs operating as active resistors. If the gain achievable with all the enhancement-mode devices is insufficient, depletion-mode loads can be employed.

Another method to enhance the load and increase the voltage gain of an input differential stage is to use current mirrors, as shown in figure (b), (c) and (d). Note that such circuits have a differential input but a single-ended output.

The input differential stage with a current mirror as a load also provides a higher CMRR than the symmetrical configuration shown in figure (a). The simplest way to substantially increase the small-signal load resistance is to employ a Wilson or cascode current mirror. To reduce the Miller effect and increase the output resistance we can use a cascode load, as shown in figure (d).

In the circuit, the common gate devices M3, M4 isolate the input and output and therefore, prevent the Miller effect. A common disadvantage of the differential stages shown in figure (c) and (d) is the reduced output voltage swing due to many cascaded devices.

The above circuits can be modified to other differential amplifiers called folded-cascode configurations, as shown in figure (a) and (b). Such amplifiers are characterized by a wide bandwidth and increased PSRR, high gain and a wide input common-mode range. Therefore, they are suitable for high frequency applications. It should be noted that to provide high-swing capability for the output voltage an improved high-swing n-channel cascode current mirror has been used.

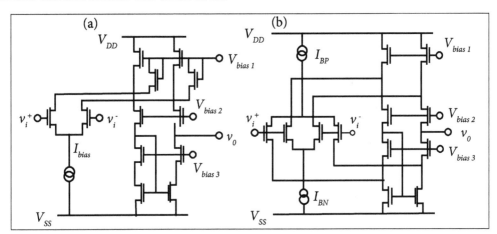

Folded-cascode differential stage: (a) With NMOS input devices and with slew-rate enhancement and (b) Symmetrical with NMOS and PMOS input devices.

2.5.1 Two Stage Amplifier

Often two or three gain stages are used in order to realize an appropriately high-gain operational amplifier. More than three stages are rather seldom used since stability problems may occur. By connecting the basic sub-circuits, we can construct different configurations of MOS op-amps.

Standard CMOS two-stages op-amps with p-channel and n-channel MOSFETs used as input stages are shown in figure (a) and (b) respectively. Differential gain stages are used as input stages and well-known inverter stages are used as second amplifying stages. To provide an appropriate frequency characteristics, a simple RC branch is connected between output of the first and second stage, which is called as the frequency compensation network.

The op-amps shown in figure (a) and (b) can be modeled for small-signals in a wide frequency range by the circuit shown in figure (c) where R_j and C_j $(j = 1, 2)$ are the output resistances and capacitances of the j-th stage, respectively. The capacitor C2 at the output node of the second stage includes the capacitive load CL and the parasitic capacitances of the second stage C_s.

Note that the dc voltage gain of the j-th stage is given by,

$$A_{ovj} = g_j R_j (j = 1, 2)$$ ...(1)

The overall dc voltage gain can be easily determined from the circuit (by simply removing all capacitors) as,

$$A_o = A_{o1}A_{o2} := \frac{V_o}{v_i^+ - v_i^-} \cong g_1 g_2 R_1 R_2 = \frac{g_{m1}}{g_{d1} + g_{d3}} \frac{g_{m7}}{g_{d6} + g_{d7}}$$ ...(2)

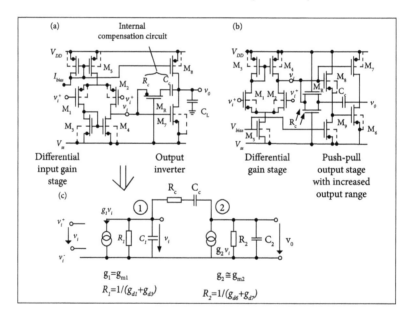

Basic CMOS op-amps with internal compensation: (a) op-amp configuration with an input differential gain stage using input PMOS devices and a simple inverter as a second gain stage, (b) op-amp configuration with gain stage using input NMOS devices and an improved, class AB, push-pull output stage and (c) simplified model of the two-stage CMOS op-amps.

To find the voltage gain for the high frequencies, we apply KCL to nodes 1 and 2 of the circuit shown in figure (c).

$$g_1 V_i + V_1/R_1 + V_1 sC_1 + (V_1 - V_0)/(R_c + 1/sC_c) = 0$$

$$g_2 V_1 - (V_1 - V_0)/(R_c + 1/sC_c) + V_0/R_2 + V_0 sC_2 = 0$$ ...(3)

Solving these equations with respect to the output voltage Vo and taking into account that in practice the conditions $A_o >> 1$, $R_c << R_1$, $R_c << R_2$, $C_1 << C_c$, $C_1 << C_L$ and $C_2 << C_L$ are fulfilled, we obtain after some manipulations,

$$A(s) \frac{V_o(s)}{V_i(s)} \cong \frac{A_o(1-s/s_z)}{(1-s/p_1)(1-s/p_2)(1-s/p_3)} \qquad ...(4)$$

Where,

$$s_z \cong -\left[C_c\left(R_c - 1/g_2\right)\right]^{-1}$$

$$p_1 \cong \frac{-1}{g_2 R_1 R_2 C_c} \cong \frac{-g_1}{A_o C_c}$$

$$p_2 \cong \frac{-g_2 C_c}{C_1 C_2 + C_c(C_1 + C_2)} \cong \frac{-g_2}{C_L}, \qquad p_3 \cong -\frac{1}{C_1 R_c}$$

Note that $|p_1| < |p_2| < |p_3|$

To provide a good closed-loop stability of the op-amp, only the dominant pole p_1 must be below the unity-gain frequency ω_t. The main task in compensating the op-amp is to move all the poles and zero, except for the dominant pole (p_1), sufficiently beyond the unity-gain frequency ω_t. i.e., to create an approximate single pole frequency response model of the op-amp.

We have two possibilities for creating an approximate single pole frequency response model of the op-amp. First, if the load capacitance C_L is large, it may be convenient to use the pole/zero cancellation technique, i.e., the resistance R_c should be chosen so that the zero $-s_z$ and the pole -p2 have the same values. This may be satisfied if,

$$R_c = \frac{C_c + C_L}{g_2 C_c} \qquad ...(5)$$

The resulting cancellation ensures that the op-amp is described by a two-pole model. However, for practical values of the parameters, we have $|p_1| << |p_3|$ and the non-dominant pole p_3 can be neglected.

Note that in such a case, the gain-bandwidth product of the op-amp can be expressed as,

$$GB = A_o |p_1| = g1/C_c$$

Thus for unity-gain stability, it is required that,

$$|p_3| > GB = \frac{g_1}{C_c} \qquad \text{or} \qquad \frac{1}{C_1 R_c} > \frac{g_1}{C_c}$$

Hence, considering (5) and assuming $C_L >> C_c$, we get

$$C_c > \sqrt{\frac{g_1}{g_2} C_1 C L} \qquad\qquad ...(6)$$

When the load capacitance is relatively low, it may be impractical to use the pole/zero cancellation technique. In such a case, the zero -sz is canceled by setting $R_c = 1/g_2$. To provide sufficiently wide spacing between the poles (i.e., $|p_1| << |p_2| << |p_3|$), the compensation capacitance must fulfill the condition,

$$C_c >> C_2 g_1 / g_2 \qquad\qquad ...(7)$$

and C_1 must be sufficiently small, so that,

$$C_1 << C_2 / g_2 R_c \qquad\qquad ...(8)$$

Then the poles p_2 and p_3 are placed far beyond the nominal unity-gain frequency ω_t and in standard applications, they can be neglected. In some applications, only the capacitor C_c is used in the compensation branch, i.e., $R_c = 0$.

The purpose of the compensation capacitor C_c is to move the dominant pole p1 to a much lower frequency and the second pole p2 to a much higher frequency. For this reason, the capacitor C_c is referred as a pole-splitting capacitor.

2.5.2 Frequency Response and Compensation

The compensation capacitor C_c has caused the magnitude of the gain to decrease, but still at frequencies well below the unity-gain frequency of the op-amp. This corresponds to the mid band frequencies for many applications. This allows us to make a couple of simplifying assumptions.

First, we will ignore all the capacitors except the compensation capacitor, C_c, which normally dominates at all the frequencies except around the unity-gain frequency of the op-amp.

Then, we also assume that one of the transistor is not present. This transistor operates as a resistor, which is included to achieve the lead compensation and it has an effect only around the unity-gain frequency of the op-amp. The simplified circuit used for analysis is shown in the below figure.

It is worth mentioning that this simplified circuit is often used during the system-level simulations when speed is more important than accuracy, but aspects such as slew-rate limiting should be included in the simulation.

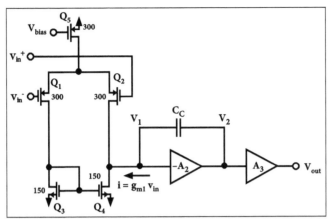

The simplified model for the opomp used to find the midband frequency response.

The second stage introduces a capacitive load on the first stage due to the compensation capacitor, C_c. Using Miller's Theorem, we can show that the equivalent load capacitance, C_{eq} at node v1 is given by,

$$C_{eq} = C_c \left(1 + A_2\right) \approx C_c A_2 \qquad \qquad ...(1)$$

The gain in the first stage can now be found using the small-signal model, resulting in

$$A_1 = \frac{V_1}{V_{in}} = -g_{m1} Z_{out1}$$

Where,

$$Z_{out} = r_{ds2} \| r_{ds4} \| \frac{1}{sC_{eq}} \qquad \qquad ...(2)$$

For mid-band frequencies, the impedance of C_{eq} dominates, and we can write

$$Z_{out1} \cong \frac{1}{sC_{eq}} \cong \frac{1}{sC_c A_2} \quad ...(3)$$

For the overall gain we have

$$A_v\left(s\right) \equiv \frac{V_{out}}{V_{In}} = A_3 A_2 A_1 \cong A_3 A_2 \frac{g_{m1}}{sC_c A_2} \qquad \qquad ...(4)$$

Using (2) and (3). If we further assume that $A_3 \cong 1$, then the overall gain, given in the above equation is simplified as,

$$A_v\left(s\right) = \frac{g_{m1}}{sC_c} \qquad \qquad ...(5)$$

This simple equation can be used to find the approximate unity-gain frequency. Specifically, to find the unity-gain frequency, ω_{ta}, we set $\left| A_v\left(j\omega_{ta}\right) \right| = 1$, and solve for ω_{ta}. Performing such a procedure with (5), we obtain the following relationship:

$$\omega_{ta} = \frac{g_{m1}}{C_c} \qquad \qquad ...(6)$$

Note here that the unity-gain frequency is directly proportional to g_{m1} and inversely proportional to C_c.

2.5.3 Slew Rate

Another important high-frequency parameter of an op-amp is its slew rate. The slew rate is the maximum rate at which the output changes when the input signals are large. When the op-amp is limited by its slew rate because a large input signal is present, all of the bias current of Q_5 goes into either Q_1 or Q_2, depending on whether V_{in} is negative or positive.

When V_{in} is a large positive voltage, the bias current I_{D5}, goes entirely through Q_1 and also goes into the current-mirror pair Q_3, Q_4. Thus, the current coming out of the compensation capacitor, C_c (i.e., I_{D4}) is simply equal to I_{D5} since Q_2 is off.

When Vin is a large negative voltage, the current-mirror pair Q_3 and Q_4 is shut off because Q_1 is off and now the bias current I_{D5}, goes directly into. In either case, the maximum current entering or leaving C_c is simply the total bias current I_{D5}.

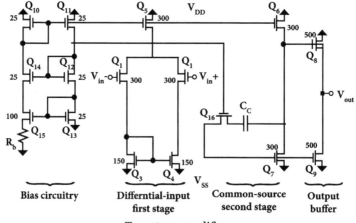

Two-stage amplifier.

Defining the slew rate, SR to be the maximum rate that V_2 can change and recalling that $V_{out} \cong V_2$, we have,

$$SR \left. \frac{d\,v_{out}}{dt} \right|_{max} = \left. \frac{I_{C_c}}{C_c} \right|_{max} = \frac{I_{D5}}{C_c} \qquad \qquad ...(7)$$

Where we used the charge equation $q = CV$, which leads to $I = dq/dt = C(dV/dt)$. Since $I_{D5} = 2I_{D1}$, we can also write,

$$SR = \frac{2I_{DI}}{C_c} \qquad \text{...(8)}$$

Where I_{D1} is the original bias current of Q1 with no signals present. Also using (6), we have $C_c = g_{m1}/\omega_{ta}$ and substituting this into (8), we have,

$$SR = \frac{2I_{DI}\omega_{ta}}{g_{m1}} \qquad \text{...(9)}$$

Recalling that,

$$g_{m1} = \sqrt{2\mu_p C_{ox}\left(\frac{W}{L}\right)I_{D1}} \qquad \text{...(10)}$$

We finally have another relationship for the slew-rate value.

$$SR = \frac{2I_{D1}}{\sqrt{2\mu_p C_{ox}(W/L)_1 I_{D1}}}\omega_{ta} = V_{eff1}\omega_{ta}$$

Where,

$$V_{eff1} = \sqrt{\frac{2I_{D1}}{\mu_p C_{ox}(W/L)_1}}$$

Normally we has little control over ω_{ta}, assuming a given maximum power dissipation is allowed. It is usually constrained to be less than two-thirds of the second-pole frequency on compensation. As a result, the only way of improving the slew rate for the two-stage CMOS op-amp is to increase V_{eff1}, ω_{ta} or both.

Obtaining a high slew rate and unity-gain frequency are two of the major reasons for choosing p-channel input transistors rather than n-channel input transistors. It should be mentioned here that increasing V_{eff1} lowers the trans-conductance of the input stage. Although increasing V_{eff1} helps to minimize distortion, a lower trans-conductance in the first stage decreases the dc gain and increases the equivalent input thermal noise.

3

Testing and Measurements

3.1 Overview of Mixed-Signal Testing

A mixed-signal circuit can be defined as a circuit consisting of both the digital and analog elements. By this definition, a comparator is one of the simplest mixed-signal circuits. It compares two analog voltages and determines if the first voltage is greater than or less than the second voltage. Its digital output changes to one of two states, depending on the outcome of the comparison.

In effect, a comparator is a one-bit analog-to-digital converter (ADC). It might also be argued that a simple digital inverter is a mixed-signal circuit, since its digital input controls an "analog" output that swings between two fixed voltages, rising, falling, over-shooting and undershooting according to the laws of analog circuits.

In fact, in certain extremely high-frequency applications, the outputs of digital circuits have been tested using mixed-signal testing methodologies. Some mixed-signal experts might argue that a comparator and an inverter are not mixed-signal devices at all. The comparator is typically considered as an analog circuit, while an inverter is considered as a digital circuit.

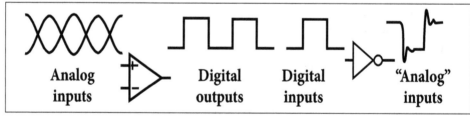

Comparator and inverter.

Other examples of borderline mixed-signal devices are analog switches and program-mable gain amplifiers. The purist might argue that mixed-signal circuits are those that involve some sort of non-trivial interaction between digital signals and analog signals. Otherwise, the device is simply a combination of digital logic and separate analog cir-cuitry coexisting on the same circuit board.

The line between the mixed-signal circuits and analog or digital circuits is blurry if we want to be pedantic. Fortunately, the blurry lines between digital, analog and mixed-sig-nal are completely irrelevant in the context of mixed-signal test and measurement. Most complex mixed-signal devices include at least some stand-alone analog circuits that do not interact with digital logic at all.

3.1.1 Test and Diagnostic Equipment

Automated Test Equipment

Automated test equipment is available from a number of commercial vendors, such as Teradyne, LTX-Credence, and Advantest, to name a few. The Teradyne, Inc. Flex mixed-signal tester is shown in the below figure. High-end ATE testers often consist of three major components such as a test head, a workstation and the mainframe. The computer workstation serves as the user interface to the tester.

The test engineer can debug the test programs from the workstation using a variety of software tools from the ATE vendor. Manufacturing personnel can also use the workstation to control the day-to-day operation of the tester as it tests the devices in production.

Teradyne Flex mixed-signal tester.

The mainframe contains power supplies, measurement instruments, and one or more computers that control the instruments as the test program is executed. The mainframe may also contain a manipulator to position the test head precisely. It may also contain a refrigeration unit to provide the cooled liquid to regulate the temperature of the test head electronics.

Although much of the tester's electronics are contained in the mainframe section, the test head contains the most sensitive measurement electronics. These circuits are the ones that require close proximity to the device under test. For example, high-speed digital signals benefit from short electrical paths between the tester's digital drivers and the pins of the DUT. Therefore, the ATE tester's digital drivers and receivers are located in the test head close to the DUT.

A device interface board (DIB) forms the electrical interface between the ATE tester and the DUT. The DIB is also known as a performance board, family board or swap block depending on the ATE vendor's terminology. DIB comes in many shapes and sizes, but their main function is to provide a temporary electrical connection between DUT and the electrical instruments in the tester.

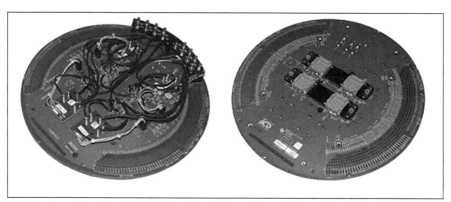

Octal site device interface board (DIB) showing DUT sockets (left) and local circuits with RF interface (right).

The DIB also provides space for DUT-specific local circuits such as load circuits and buffer amplifiers that are often required for mixed-signal device testing. The above figure illustrates the top and bottom sides of an octal site DIB. The topside shown on the left displays eight DUT sockets and the right shows the local circuits and RF interface.

Wafer Probers

Wafer probers are robotic machines that manipulate the wafers as the individual dies are tested by the ATE equipment. The prober moves the wafer underneath a set of tiny electrical probes attached to a probe card. The probes are connected to the electrical resources of the ATE tester through a probe interface board (PIB).

The PIB is a specialized type of DIB board that may be connected to the probe card through coaxial cables or spring-loaded contacts called pogo pins. The PIB and probe card serve the same purpose for the wafer that the DIB board serves for the packaged device. They provide a means of temporarily connecting the DUT to the ATE tester's electrical instrumentation while testing is performed.

The prober informs the tester when it has placed each new die against the probes of the probe card. The ATE tester then executes a series of electrical tests on the die before instructing the prober to move to the next die. The handshaking between the tester and prober insures that the tester only begins testing when a die is in position and that the prober does not move the wafer in the midtest. Figure below shows a wafer prober and close-up views of a probe card and its probe tips.

Wafer prober and probe card.

Handlers

Handlers are used to manipulate packaged devices in much the same way that probers are used to manipulate wafers. Most handlers fall into two categories such as gravity-fed and robotic. Robotic handlers are also known as pick-and-place handlers. Gravity-fed handlers are normally used with the dual inline packages, while the robotic handlers are used with devices having pins on all four sides or pins on the underside.

Either type of handler has one main purpose. i.e., to make a temporary electrical connection between the DUT pins and the DIB board. Gravity-fed handlers often perform this task using a contactor assembly that grabs the device pins from either side with metallic contacts that are in turn connected to the DIB board. Robotic handlers usually pick up each device with a suction arm and then plunge the device into a socket on the DIB board.

In addition to provide a temporary connection to the DUT, handlers are also responsible for sorting the good DUTs from the bad ones based on the test results from the ATE tester. Some handlers also provide a controlled thermal chamber where devices are allowed to "soak" for a few minutes so they can either be cooled or heated before testing. Since many electrical parameters shift with the temperature, this is an important handler feature.

E-Beam Probers

Electron beam probers are used to probe the internal device signals while the device is being stimulated by the tester. These machines are very similar to scanning electron microscopes (SEMs). Unlike an SEM, an e-beam prober is designed to display variations in circuit voltage as the electron beam is swept across the surface of an operating DUT. Variations in the voltage levels on the metal traces in the IC appear as different shades of gray in the e-beam display. E-beam probers are extremely powerful diagnostic tools, since they provide measurement access to internal circuit nodes.

Focused Ion Beam Equipment

Focused ion beam (FIB) equipment is used in conjunction with the e-beam probers to modify the device's metal traces and other physical structures. A FIB machine can cut holes in oxide and metal traces and can also lay down new metallic traces on the surface of the device. Experimental design changes can be implemented without waiting for a complete semiconductor fabrication cycle. The results of the experimental changes can then be observed on the ATE tester to determine the success or failure of the experimental circuit modifications.

Forced-Temperature Systems

The handler's thermal chamber allows characterization and testing of large numbers of DUTs at a controlled temperature. When characterizing a small number of DUTs at a variety of temperatures, a less expensive and cumbersome method of temperature control is needed. Portable forced-temperature systems allow DUT performance characterization under a variety of controlled thermal conditions.

The nozzle of a forced-temperature system can be seated against the DIB board or bench characterization board, forming a small thermal chamber for the DUT. Many forced-temperature systems are able to raise or lower the DUT's ambient temperature across the full military range (-55 to +125°C).

3.1.2 Mixed-Signal Testing Challenges

Time to Market

Time to market is a pressing issue for the semiconductor manufacturers. Profit margins for a new product are highest shortly after the product has been released to the market. Margins begin to shrink as competitors introduce similar products at lower prices. The lack of a complete, cost-effective test program can be one of the bottlenecks preventing the release of a new product to profitable volume production.

Mixed-signal test programs are particularly difficult to produce in a short period of time. Surprisingly, the time spent writing test code is often significantly less than the time spent learning about the device under test, designing the test hardware, defining the test plan and debugging the ATE test solution once silicon is available.

Much of the time spent in the debugging phase of test development is actually spent debugging the device problems. Mixed-signal test engineers often spend as much time running experiments for design engineers to isolate the design errors as they spend debugging their own test code. Perhaps the most aggravating debug time of all is the time spent tracking down the problems with the tester itself or the tester's software.

Accuracy, Repeatability and Correlation

Accuracy is a major concern for mixed-signal test engineers. Inaccurate answers are caused by a bewildering the number of problems. Improperly calibrated instruments, electromagnetic interference, improperly ranged instruments and measurements made under incorrect test conditions can all lead to inaccurate test results.

Repeatability is the ability of the test equipment and test program to give the same answer multiple times. Actually, a measurement that never changes at all is suspicious. It sometimes indicates that the tester is improperly configured, giving the same incorrect answer repeatedly. A good measurement typically shows some variability from one test program execution to the next, since electrical noise is present in all the electronic circuits. Electrical noise is the source of many repeatability problems.

Another problem facing mixed-signal test engineers is correlation between the answers given by different pieces of measurement hardware. The design engineer often finds that the test program results do not agree with the measurements taken using bench equipment in their lab.

The test engineer must determine which answer is correct and why there is a discrepancy. It is also common to find that two supposedly identical testers or DIB boards give different answers or that the same tester gives different answers from day to day. These problems frequently result from obscure hardware or software errors that may take days to isolate. Correlation efforts can represent a major portion of the time spent debugging a test program.

Electromechanical Fixturing Challenges

The test head and DIB board must ultimately make contact to the DUT through the handler or prober. There are few mechanical standards in the ATE industry to specify how a tester should be docked to a handler or prober. The test engineer has to design a DIB board that not only meets electrical requirements, but also meets the mechanical docking requirements. These requirements include connector locations, board thickness, DUT socket mechanical holes and various alignment pins and holes.

Handlers and probers must make a reliable electrical connection between the DUT and the tester. Unfortunately, the metallic contacts between DUT and DIB board are often very inductive and capacitive. Stray inductance and capacitance of the contacts can represent a major problem, especially when testing the high-impedance or high-frequency circuits.

Although several companies have marketed test sockets that reduce these problems, a socketed device will often not perform quite as well as a device soldered directly to a printed circuit board. Performance differences due to sockets are yet another potential source of correlation error and extended time to market.

Economics of Production Testing

Time is money, especially when it comes to the production test programs. A high-performance tester may cost two million dollars or more, depending on its configuration. For a specific class of devices developed with the design-for-test in mind, the test system can be reduced to as low as two hundred thousand dollars.

Probers and handlers may cost five hundred thousand dollars or more. If we also include the cost of providing floor space, electricity and production personnel, it is easy to understand why testing is an expensive business.

One second of test time can cost a semiconductor manufacturer one to six cents. This may not seem expensive at first glance, but when test costs are multiplied by millions of devices a year, the numbers add up quickly. For example, a five-second test program costing four cents per second times one million devices per quarter costs a company eight hundred thousand dollars per year in bottom-line profit.

Testing can become the fastest-growing portion of the cost of manufacturing a mixed-signal device if the test engineer does not work on the cost optimized test solution. When testing instead of one device at a time, the test solution allows testing eight devices or more with one test insertion, the test cost can be kept under control to be still in the single-digit percentage of the build cost of the device.

Continuous process improvements and better photolithography allow the design engineers to add more functions on a single semiconductor chip at little or no additional cost. Unfortunately, test time cannot be similarly reduced by simple photolithography. A 100-Hz sine wave takes 10 ms per cycle no matter how small we shrink a transistor.

One feature common in ATE is multisite capability. Multisite testing is a process in which multiple devices are tested on the same test head simultaneously with obvious savings in test cost. The word "site" refers to each socketed DUT. For example, site 0 corresponds to the first DUT, site 1 corresponds to the second DUT, and so on. Multisite testing is primarily a tester operating system feature, although the duplicate tester instruments must be added to the tester to allow simultaneous testing on multiple DUT sites.

Clearly, production test economics is an extremely important issue in the field of mixed-signal test engineering. Not only must, the test engineer perform accurate measurements of mixed-signal parameters, but the measurements must be performed as quickly as possible to reduce the production costs. Since a mixed-signal test program may perform hundreds or even thousands of measurements on each DUT, each measurement must be performed in a small fraction of a second.

3.2   Test Specification Process: Device Datasheets

A test plan is a written list of tests and test procedures that will be used to verify the quality of a particular device under test (DUT). The definition of a production test plan usually begins with a device data sheet or specification sheet, as it is often called.

Unfortunately, the data sheet does not directly translate into a finite list of all required production tests. For example, a low-pass filter ripple specification of ±1.0dB states that gain variation is guaranteed at each and every frequency in the pass band of the filter. Semiconductor manufacturers do not test every possible frequency in production.

Test plan generation sometimes seems like more of an art than a science, especially when one tries to define exactly how a data sheet is translated into a test plan.

The data sheet serves many purposes. When development of a new device begins, the datasheet serves as the design specification. Design engineers refer to the data sheet as a blueprint to make sure they design the functions that the marketing and systems engineering organizations have specified.

As the project progresses, the test and product engineers refer to the data sheet to define the test list. The test list must be comprehensive enough to guarantee that the manufactured devices meet the data sheet specifications. Throughout the design process, the customer refers to the data sheet while designing the device into the system level end application.

The data sheet thus serves as the formal communication channel between the marketing and engineering personnel engaged in a project. The test engineer often detects data sheet mistakes and ambiguities while writing the test plan or developing the test program. In effect, the test engineer is the first customer for a new design.

Likewise, the tester and device interface board (DIB) can be considered the new device's first application. Data sheet errors should be promptly corrected to prevent further mistakes. For instance, if an inappropriate measurement is specified in the initial data sheet, it is the test engineer's responsibility to make sure the error gets corrected or clarified so that a sensible test plan can be defined. For this reason, it is important to know which organization is responsible for controlling the data sheet's contents.

It must meet that customer's exact requirements. In the case of a catalog device, the systems engineering or marketing organization controls the data sheet. The test engineer only needs to get agreement from the design and systems engineers to make a data sheet change. In the case of a custom device, the customer and systems engineer share responsibility for the contents of the data sheet. In addition to approvals from the marketing or systems engineering team, the customer's approval is also required before the data sheet can be modified.

Depending on the customer's requirements, data sheet changes may be very easy to implement or they may be impossibly difficult. Regardless of the customer's needs, though specification changes requested at the last minute give the appearance of a poorly run organization. For this reason, it is a good idea for the test engineer to get involved very early in the definition of a device so that specification changes can be suggested in a timely manner. Suggestions made early in the new product development cycle give a customer more confidence that the testing process is under control.

Structure of a Data Sheet

Data sheets may contain any of the following sections:

- A feature summary and description.

- Principles of operation.

- Absolute maximum ratings.

- Electrical characteristics.

- Timing diagrams.

- Application information.

- Characterization data.

- Circuit schematic and die layout.

Figure below shows an example data sheet for a digital-to-analog converter (DAC). This datasheet is taken from a Texas Instruments data acquisition circuit's data book. The first page of the data sheet provides a quick device summary. The feature summary allows the customer to quickly gauge the device's fit to a particular application.

The test engineer can generally ignore this section since the same information is typically called out in subsequent sections of the data sheet. The pinout and package information is much more relevant to test engineering. The test engineer refers to the pinout and package information to design the DIB for each package type.

The device description gives a quick overview of the device's functionality. Together with the principles of operation, the device description defines the various operations of the device in detail. The test program must guarantee all these functions, though not necessarily in a straightforward manner. For instance, the device description may depict a circuit that divides an externally generated 1-MHz reference clock by one million, producing a 1-s time base.

Since straightforward testing would represent an unacceptably long test time of 1 s, this function might be tested in an indirect manner. There are many indirect ways to guarantee the operation of a 1-s time base counter without spending 1s of test time. There are two kinds of devices such as catalog and custom. A catalog device is one that is defined by the semiconductor manufacturer or by an IC design house. Once defined, a catalog need to split the divided into two halves and may not even need to know that it can be placed in this test mode at all. Therefore, test modes may or may not be documented in the data sheet.

Data sheet for 8-bit multiplying DAC: Features, description and pinout

- Easily Interfaced to Microprocessors.

- On-Chip Data Latches.

- Monotonic Over the Entire A/D Conversion Range.

- Segmented High-Order Bits Ensure Low-Glitch Output.

- Interchangeable With Analog Devices AD7524, PMI PM-7524, and Micro Power Systems MP7524.

- Fast Control Signaling for Digital Signal-Processor Applications Including Interface With TMS320.

- CMOS Technology.

Key Performance Specifications	
Resolution	8 Bits
Linearity error	1/2 LSB Max
Power dissipation at VDD = 5 V	5 mW Max
Setting time	100 ns Max
Propagation delay time	80 ns Max

Description

The TLC7524C, TLC7524E, and TLC7524I are CMOS, 8-bit, digital-to-analog converters (DACs) designed for easy interface to most popular microprocessors.

The devices are 8-bit, multiplying DACs with input latches and load cycles similar to the write cycles of a random access memory. Segmenting the high-order bits minimizes glitches during changes in the most significant bits, which produce the highest glitch impulse. The devices provide accuracy to 1/2 LSB without the need for thin-film resistors or laser trimming, while dissipating less than 5 mW typically.

Featuring operation from a 5-V to 15-V single supply, these devices interface easily to most microprocessor buses or output ports. The 2- or 4-quadrant multiplying makes these devices an ideal choice for many microprocessor-controlled gain-setting and signal-control applications.

The TLC7524C is characterized for operation from 0°C to 70°C. The TLC7524I is characterized for operation from −25°C to 85°C. The TLC7524E is characterized for operation from − 40°C to 85°C.

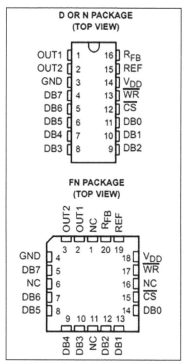

NC–No internal connection.

AVAILABLE OPTIONS			
TA	PACKAGE		
	SMALL OUTLINE PLASTIC DIP (D)	PLASTIC CHIP CARRIER (FN)	PLASTIC DIP (N)
0°C to 70°C	TLC7524CD	TLC7524CFN	TLC7524CN
−25°C to 85°C	TLC7524ID	TLC7524IFN	TLC7524IN
−40°C to 85°C	TLC7524ED	TLC7524EFN	TLC7524EN

Eight-bit Multiplying DAC: Principles of Operation

Principles of Operation

The TLC7524C, TLC7524E, and TLC7524I are 8-bit multiplying DACs consisting of an inverted R-2R ladder, analog switches, and data input latches. Binary-weighted currents are switched between the OUT1 and OUT2 bus lines, thus maintaining a constant current in each ladder leg independent of the switch state. The high-order bits are decoded. These decoded bits, through a modification in the R-2R ladder, control three equally-weighted current sources. Most applications only require the addition of an external operational amplifier and a voltage reference.

The equivalent circuit for all digital inputs low is seen in Figure 2. With all digital inputs low, the entire reference current, Iref, is switched to OUT2. The current source I/256 represents the constant current flowing through the termination resistor of the R-2R ladder, while the current source Ilkg represents leakage currents to the substrate. The capacitances appearing at OUT1 and OUT2 are dependent upon the digital input code. With all digital inputs high, the off-state switch capacitance (30 pF maximum) appears at OUT2 and the on-state switch capacitance (120 pF maximum) appears at OUT1. With all digital inputs low, the situation is reversed as shown in Figure 2. Analysis of the circuit for all digital inputs high is similar to Figure 2; however in this case, Iref would be switched to OUT1.

The DAC on these devices interfaces to a microprocessor through the data bus and the $\overline{\text{CS}}$ and $\overline{\text{WR}}$ control signals. When $\overline{\text{CS}}$ and $\overline{\text{WR}}$ are both low, analog output on these devices responds to the data activity on the DB0–DB7 data bus inputs. In this mode, the input latches are transparent and input data directly affects the analog output. When either the $\overline{\text{CS}}$ signal or $\overline{\text{WR}}$ signal goes high, the data on the DB0–DB7 inputs are latched until the $\overline{\text{CS}}$ and $\overline{\text{WR}}$ signals go low again. When CS is high, the data inputs are disabled regardless of the state of the WR signal.

These devices are capable of performing 2-quadrant or full 4-quadrant multiplication. Circuit configurations for 2-quadrant or 4-quadrant multiplication are shown in Figures 3 and 4. Tables 1 and 2 summarize input coding for unipolar and bipolar operation respectively.

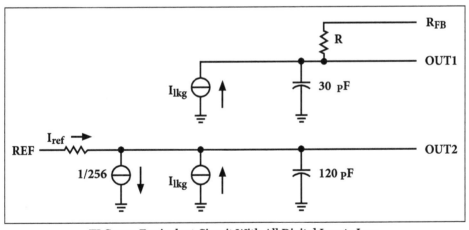

TLC7524 Equivalent Circuit With All Digital Inputs Low.

Many of the features listed in the device description and principles of operation do not results in the measurements of electrical parameters. These features are verified using a go/no-go test or functional test. Functional tests result in a simple pass/fail result with no numerical reading. Parametric tests, by comparison are those that return a value that must be compared against one or more test limits to determine pass/fail results.

The 1-s timer is a good example of a circuit that can be tested with a functional test. It is not necessary to measure the exact countdown period in seconds and fractions of a second. The digital counter circuit either divides by one million or it does not. This type of digital logic verification is known as a functional pattern test.

The only way the 1-s period of time could be in error is if the divider circuits are not functional or if the 1-MHz external reference clock is not set to the correct frequency. An incorrect external frequency setting does not need to be tested during IC production, since it is not a function of device performance. Only a functional pattern test is required to guarantee the 1-s interval.

An automated software process is often used to generate functional pattern tests. The test engineer should verify that all digital functionality has been guaranteed by either automatically generated patterns or by hand-coded functional pattern tests.

In highly customized mixed-signal devices, the test engineer needs to understand the end application of the device. The device description and principles of operation provide the test engineer with much of the information needed to understand the system into which the device will be placed.

Electrical Characteristics

Electrical characteristics (or electrical specifications) provide the test limits and test conditions for many of the parametric tests in a mixed-signal test program. Figure below show the electrical characteristics for the 8-bit multiplying DAC.

Recommended operating conditions and electrical characteristics

MIN		$V_{DD} = 5$ V		MIN	$V_{DD} = 15$ V		UNIT
		NOM	MAX		NOM	MAX	
Supply voltage, V_{DD}		4.75	5	5.25	14.5	15	15.5	V
Reference voltage, V_{ref}			±10			±10		V
High-level input voltage, V_{IH}		2.4						V
Low-level input voltage, V_{IL}				0.8			1.5	V
CS setup time, $t_{su(CS)}$		40			40			ns
CS hold time $t_{h(CS)}$		0			0			ns
Data bus input setup time $t_{su(D)}$		25			25			ns
Data bus input hold time $t_{h(D)}$		10			10			ns
Pulse duration, $\overline{WR}$ low, $t_{w(WR)}$		40			40			ns
Operating free-air temperature, TA	TLC7524C	0		70	0		70	°C
	TLC7524I	-25		85	-25		85	
	TLC7524E	-40		85	-40		85	

Electrical characteristics over recommended operating free-air temperature range, Vref = ± 10 V, OUT1 and OUT2 at GND (unless otherwise noted).

PARAMETER		TEST CONDITIONS	V_{DD} = 5V			V_{DD} = 15V			UNIT
			MIN	TYP	MAX	MIN	TYP	MAX	
I_{IH} High-level input current		$V_I = V_{DD}$			10			10	μA
I_{IL} Low-level input current		$VI = 0$			-10			-10	μA
I_{lkg} Output leakage current	OUT1	DB0–DB7 at 0 V, $\overline{WR}, \overline{CS}$ at 0 V, $V_{ref} = \pm10$ V			±400			±200	nA
	OUT2	DB0–DB7 at V_{DD}. $\overline{WR}, \overline{CS}$ at 0 V, $V_{ref} = \pm10$ V			±400			±200	
I_{DD} Supply current	Quiescent	DB0–DB7 at V_{IH} min or V_{IL} max			1			2	mA
	Standby	DB0–DB7 at 0 V or V_{DD}			500			500	μA
k_{SVS} Supply voltage sensitivity, $\Delta gain/\Delta VDD$		$\Delta V_{DD} = \pm10\%$		0.01	0.16		0.005	0.04	%FSR/%
Ci Input capacitance, DB0–DB7, $\overline{WR}$, $\overline{CS}$		$VI = 0$			5			5	pF
C_o Output capacitance	OUT1	DB0 DB7 at 0 V, $\overline{WR}, \overline{CS}$ at 0 V			30			30	pF
	OUT2	DB0 DB7 at VDD, $\overline{WR}, \overline{CS}$ at 0 V			120			120	
	OUT3				120			120	
	OUT4				30			30	
Reference input impedance (REF to GND)			5		20	5		20	kΩ

Electrical Specifications and Timing Diagram

Operating characteristics over recommended operating free-air temperature range, Vref = ±10 V, OUT1 and OUT2 at GND (unless otherwise noted).

PARAMETER	TEST CONDITIONS	V_{DD}= 5V			V_{DD} = 15 V			UNIT
		MIN	TYP	MAX	MIN	TYP†	MAX	
Linearity error				±0.5			±0.5	LSB
Gain error	See Note 1			±2.5			±2.5	LSB
Settling time (to 1/2 LSB)	See Note 2			100			100	ns
Propagation delay from digital input to 90% of final analog output current	See Note 2			80			80	ns
Feed through at OUT1 or OUT2	Vref = ±10 V (100-kHz sinewave) $\overline{WR}$ and $\overline{CS}$ at 0 V, DB0–DB7 at 0 V			0.5			0.5	%FSR
Temperature coefficient of gain	TA = 25°C to MAX		±0.004			±0.001		%FS-R/°C

NOTES:

- Gain error is measured using the internal feedback resistor. Nominal full scale range (FSR) = V_{ref} – 1 LSB.

- OUT1 load = 100 Ω, Cext = 13 pF, $\overline{WR}$ at 0 V, $\overline{CS}$ at 0 V, DB0–DB7 at 0 V to VDD or VDD to 0 V.

Operating Sequence

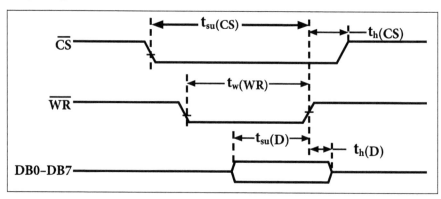

3.2.1 Generation of Test Plan

To Plan or Not to Plan

Strictly speaking, test plans are not absolutely necessary. A test engineer can certainly generate a test program by simply sitting down at the tester computer and entering code based on the device data sheet. There are several problems with this type of undisciplined approach.

First, device testability will probably not be identified early enough to allow the addition of test features to the design. Test plans force the design engineers and test engineers to work through all the details of testing at an early stage in the design cycle. Second, the test engineer may create test-to-test compatibility problems if the details of all tests are not known up front.

For example, a clocking scheme that works well for one test may be incompatible with the clocking scheme required for a subsequent test. The first test may then need to be rewritten from scratch so that the clocking schemes mesh properly.

If a test plan is not clearly documented before coding begins, then the test engineer lacks the necessary overview of the test program that allows all the tests to fit together efficiently.

Eight-bit Multiplying DAC: Application Information

Voltage-Mode Operation

It is possible to operate the current-multiplying DAC in these devices in a voltage mode. In the voltage mode, a fixed voltage is placed on the current output terminal. The analog output voltage is then available at the reference voltage terminal. Figure 1 is an example of a current-multiplying DAC, which is operated in voltage mode.

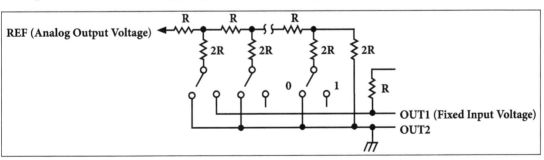

Figure 1. Voltage Mode Operation.

The relationship between the fixed-input voltage and the analog-output voltage is given by the following equation:

$$V_O = V_I (D/256)$$

Where,

V_O = analog output voltage.

V_I = fixed input voltage.

D = digital input code converted to decimal.

In voltage-mode operation, these devices meet the following specification:

PARAMETER	TEST CONDITIONS	MIN	MAX	UNIT
Linearity error at REF	VDD = 5 V, OUT1 = 2.5 V, OUT2 at GND, TA = 25°C		1	LSB

Functional Block Diagram and Absolute Maximum Rating

Functional block diagram

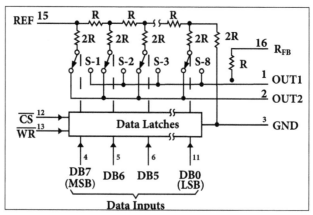

Terminal numbers shown are for the D or N package.

Absolute maximum ratings over operating free-air temperature range (unless otherwise noted).

Supply voltage range, V_{DD} . . . . −0.3 V to 16.5 V

Digital input voltage range, V_I . . . . −0.3 V to VDD + 0.3 V

Reference voltage, V_{ref} . . . . ±25 V

Peak digital input current, I_I . . . . .10 µA

Operating free-air temperature range, T_A: TLC7524C . . . . 0°C to 70°C

TLC7524I . . . . −25°C to 85°C

TLC7524E . . . . −40°C to 85°C

Storage temperature range, T_{stg} . . . . −65°C to 150°C

Case temperature for 10 seconds, T_C: FN package . . . . .260°C

Lead temperature 1, 6 mm (1/16 inch) from case for 10 seconds: D or N package 260°C

Low-Offset JFET op Amp

Typical Characteristics

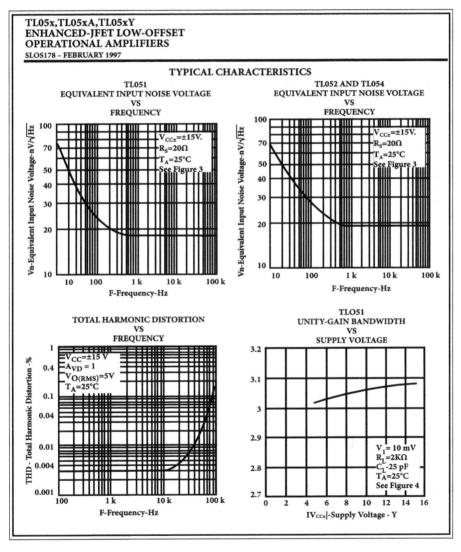

Similarly, test hardware such as DIBs and probe interface hardware cannot be properly designed until all the test details are known. Finally, the test plan helps to identify shortfalls in the target tester's capabilities. Early identification of tester deficiencies allows the test engineer time to find acceptable work-around. Sometimes, the design engineer can even modify the IC design to accommodate tester deficiencies.

3.2.2 Components of a Test Program

The test program is a detailed, tester-specific version of the test plan, written in the target tester's native language. It may at times deviate from the test plan if the target tester is incapable of performing tests exactly as specified by the test plan. In these cases, comments should be added to the program to make this discrepancy clear. Major deviations should be approved by the other members of the engineering team.

Tester languages vary from low-level C routines to very sophisticated graphical user interface environments. Despite wide differences in their details, tester languages often share some basic structural components.

Test programs typically consist of all or most of the following sections:

- Waveform creation and other tester initializations

- Calibrations

- Continuity

- DC parametric tests

- AC parametric tests

- Digital patterns (also known as functional tests)

- Digital timing tests

- Test sequence control

- Test limits

- Binning control

Well-written test programs often contain extensive characterization code to perform tests not specifically required by the data sheet. These characterization tests allow the design engineers to better evaluate the quality and robustness of the IC design.

A thorough test program may also contain code that allows offline simulation of the tests so that certain portions of the program can be debugged without a tester or device.

Test Code and Digital Patterns

Test code and digital patterns make up the bulk of mixed-signal test programs. Test code controls the order and timing of instrument settings, signal generation and signal measurements that make up each measurement in the test program. Test code typically does not control the real-time details of each instrument, though.

For example, the data generated from the digital subsystem of a tester are not clocked out one bit at a time by the test code. Instead, the test code simply calls for the tester's digital subsystem to begin exercising the desired digital pattern at the appropriate time. The digital subsystem then takes care of the details of generating the individual ones and zeros. Thus, the test code for a DAC full-scale output test might look something like this in pseudocode:

```
dac_full_scale_voltage(

Set VI1 = 2.5 V;    /* Set the DAC's voltage reference to 2.5 volts */

start digital pattern = "dac_full_scale";  /*Set the DAC output to +full scale
(2.5 V) */

connect meter: DAC_OUT    /* Connect the DAC voltmeter to the DAC output */

fsout = read_meter();     /* Read the voltage level at the DAC_OUT pin */

test fsout; /* Compare the DAC full scale output to the data sheet limits */
```

Digital patterns consist of groups of data bits called vectors. Each vector represents the drive and expect data that are to be sent out on each of the tester's digital pins at a specific time. Drive data specifies the desired state at the input to the DUT (HI, LOW or HIZ). Compare data (also called as expect data) specifies the required digital output from the device.

Vectors are usually sent out at a regular rate, called the bit cell rate. Digital patterns usually contain not only the 1/0 drive and expect data, but also the sequencing information for the vectors. The digital pattern sequencing commands allow branching, looping, and other vector sequencing operations that make the pattern more compact.

To generate a pair of clocks at two different frequencies from digital pins CLK1 and CLK2, we might write the following pseudo code pattern:

```
label  pattern control     CLK1 CLK2
START           0 /* Vector one */

                0 /* Vector two, etc

                1

      Jump START   1 /* Infinite loop */
```

This pattern would continue in an infinite loop, producing two frequencies. The CLK1 frequency would be twice as that of the CLK2 frequency. The test code and the digital pattern must operate in stepped synchronization for mixed-signal tests. It would be unfortunate in the DAC test above if the digital pattern "dac_full_scale" did not execute until 50 ms after the meter measurement had already been performed.

For this reason, mixed-signal testers include handshaking functions in both the test code and digital pattern control that allow the tester computer and digital pattern subsystem to keep in step with one another.

Another pattern issue unique to mixed-signal testing is that the pattern often must be executed at a very precise frequency. It is not acceptable to round off the period of the vector rate to the nearest nanosecond as is often done in purely digital test programs.

Binning

One of the functions of a test program is to sort each device into one of several categories called bins, depending on the outcome of the various tests. The most obvious bins are "pass" and "fail" but there are several others that might be added.

For example, a continuity test is usually inserted at the beginning of the test program. The purpose of the continuity test is to verify that all the electrical contacts between the tester and the DUT have been successfully connected. If a large percentage of devices fail the continuity test, this indicates a probable error in the tester hardware. It is therefore a good idea to use separate bins for continuity failures and data sheet failures so that the production staff can more easily recognize tester hardware problems.

Binning is not always a pass/fail operation. Sometimes there are different grades of passing devices. If a device is designed to operate at 100 MHz but some of the manufactured devices are actually able to operate at 120 MHz, then the test program might be set up to split these devices into two quality grades, "good" and "great".

Bin 1 might represent the 120-MHz devices, while Bin 2 might represent the devices that could only operate up to 100 MHz. The-120 MHz devices would be labeled differently than the 100-MHz devices. They would also be priced differently. We are all familiar with higher prices for faster PC microprocessors and memory chips.

Fast binning is a term used to describe a tester's ability to bin a bad device as soon as it fails any test. This is done to prevent a bad device from wasting valuable tester time after it has already produced a failing result. The test and product engineers should work together to ensure that the most commonly failed tests are placed near the beginning of the test program. This allows the tester to sort bad devices as quickly as possible.

The tester generates a binning signal that tells the handler or prober what to do with the various categories of devices. Until recent years, bad die on a wafer were often squirted with red ink dots to designate them as failures. Now this inking is commonly performed offline or is done in a purely virtual manner using pass/fail databases and production lot ill numbers. At final test, different grades of packaged devices are sorted into separate plastic tubes or trays by the handler.

Test Sequence Control

Test sequence may be controlled in a number of ways, depending on the sophistication of the tester's software environment. In older testers, the order of the various tests was simply determined by the order of test routine execution. Comparisons of measured results against test limits were performed after each measurement.

Test limits were therefore scattered all through the test program along with the instrument setups and measurement code. Such a scattered arrangement made it difficult to identify which test limits were applied to a device in a given version of the test program. This made it difficult to verify that the test program limits matched the data sheet limits.

As tester software environment is matured, a new type of test code module evolved to allow a more convenient summary of test flow, test limits, and binning information. The new code module, called a sequencer by some vendors, contains the test routine function calls, the test limits and the binning information for each test result. The sequencer code allows the programmer to order the tests and group all the test limits into a central location in the test code separate from the test routines themselves.

The sequencer code thus provides a convenient summary of the test list, test order, and pass/fail limits for each test. This makes it easier to audit the program for compliance with data sheet test limits. Depending on the tester's software environment, the sequencer modules may be coded as text or they may consist of graphical interface objects linked together with arrows to indicate program flow and binning decisions.

Waveform Calculations and Other Initializations

Mixed-signal test programs use many precomputed waveforms. A 1-kHz gain test requires a sinusoidal waveform that does not change from one program execution to the next. Waveforms that do not need to change are precomputed and stored either in arrays or directly into memory banks in the tester instruments themselves.

Digital waveforms are also precomputed and stored in the digital subsystem of the tester. Many of the required initializations such as waveform computations are performed only once when the test program is first loaded. Performing these initializations only once saves a large amount of test execution time.

Other operations, such as resetting tester instruments to a default state, must be performed each time the program is run. The details of initializations are very specific to each tester, but most testers involve some type of first-run initialization code. One major class of first-run code is focused calibrations and checkers.

Focused Calibrations and DIB Checkers

Sometimes the instrumentation in a tester does not have sufficient accuracy for a given test. If not, a special routine called a focused calibration is required when the program first runs. The focused calibration routine determines the inaccuracy of the instrument using slower, more accurate instrumentation as a reference. The inaccuracies of the faster instrument can then be corrected in a process known as software calibration.

Software calibrations must also be performed on circuitry placed on the device interface board. Assume an op amp voltage follower is placed on the DIB to buffer a weak device output. The gain and offset of the voltage follower adds errors into any measured results. The test engineer must calibrate the gain and offset of the voltage follower using focused calibrations to achieve maximum accuracy. Sometimes focused calibrations can be as difficult to develop as the device measurements themselves, especially when extreme accuracy is required in the final test.

Fortunately, many software calibrations are hidden from the user in the tester's operating system. These calibrations are performed automatically when the program is first loaded. Other calibrations are performed on a regular basis, such as once per week. Electromechanical relays, op amp circuits, comparators, and other active circuits are commonly placed on the DIB to extend the tester's functionality and accuracy. These circuits are subject to failure.

The test program should include DIB checker code to verify the functionality of any circuitry placed on the device interface board. This allows the production personnel to avoid running thousands of good devices through a bad DIB before discovering the error. DIB checker routines are usually run along with focused calibrations when the program is first loaded.

Characterization Code

Characterization tests are often added to a test program to allow thorough evaluation of the first few lots of production material. Thorough characterization of a new device is critical, since it allows the design engineers to identify and correct the marginal portions of the design.

An example of a characterization test would be a filter response test implemented at each frequency from 100 Hz to 10 kHz in 100 Hz increments. Such a test would never be cost-effective in a production test program, but it would provide thorough information about the filter's gain versus frequency characteristics.

Simulation Code

Simulation code is sometimes added to a mixed-signal test program to allow some of the mathematical routines to be verified. For example, the ideal output of a DAC might be simulated and stored into an array for use by a DAC linearity calculation routine. Offline code debugging techniques like this are a good way to reduce debug time and avoid wasting valuable tester time. However, such simulations are not entirely effective in uncovering errors such as incorrect DUT register settings or improper tester instrument range settings.

A more advanced type of simulation, known as test simulation or virtual test allows true closed-loop simulation of the tester and device. Using test simulation, a software model of the tester stimulates a model of the DUT according to the instructions in the test program. The tester model and tester operating system then capture the responses from the DUT model and compare them to test limits.

Debuggability

The three most important things in test program structure are debuggability, debuggability and debuggability. A study at Texas Instruments showed that the test program debugging process takes about 20% of an average test engineer's week. The debugging time was found to be roughly twice the time spent writing test code.

Debugging is not only a matter of finding and fixing test code bugs. It is also a matter of locating measurement correlation errors, intermittent failures, and hardware problems including bad DIE layout and broken tester modules. More important, test debugging often turns into design debugging.

Design debugging activities account for a large portion of the test program debugging time. One of the mixed-signal test engineer's most valuable roles is to help the design engineers isolate design problems. A good test engineer with a well-structured test program can quickly modify the program and run experiments for design engineers or customers.

These experiments are critical to reducing the time it takes to get the problems worked out of a new mixed-signal design. The success or failure of a mixed-signal product often depends on how well the design engineers, test engineers and product engineers work together to resolve design problems.

3.3   DC and Parametric Measurements

Purpose of Continuity Testing

Before a test program can evaluate the quality of a device under test (DUT), the DUT must be connected to the ATE tester using a test fixture such as a device interface board (DIB). The typical inter-connection scheme is shown in the below figure. When packaged devices are tested, a socket or handler contactor assembly provides the contact between the DUT and the DIB.

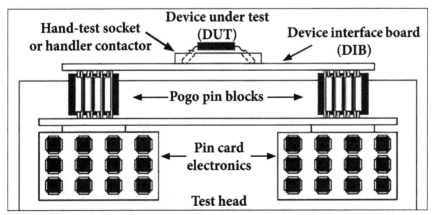

ATE test head to DUT interconnections.

When testing a bare die on a wafer, the contact is made through the probe needles of the probe card. The tester's instruments are connected to the DIB through one or more layers of connectors such as spring-loaded pogo pins or edge connectors. The exact connection scheme varies from tester to tester, depending on the mechanical or electrical performance trade-offs made by the ATE vendor.

In addition to pogo pins and other connectors, electromechanical relays are often used to route the signals from the tester electronics to the DUT. A relay is an electrical switch whose position is controlled by an electromagnetic field. The field is created by a current forced through a coil of wire inside the relay. Relays are used extensively in mixed-signal testing to modify the electrical connections to and from the DUT as the test program progresses from test to test.

Any of the electrical connections between a DUT and the tester can be defective, resulting in open circuits or shorts between electrical signals. For example, the wiper of a relay can become stuck in either the open or closed position after millions of open or close cycles. While interconnect problems may not pose a serious problem in a lab environment, defective connections can be a major source of tester down time on the production floor. Continuity tests are performed on the device to verify that all the electrical connections are sound.

If continuity testing is not performed, then the production floor personnel cannot distinguish between bad lots of silicon and defective test hardware connections. Without continuity testing, thousands of good devices could be rejected simply because a pogo pin was bent or because a relay was defective.

Continuity Test Technique

Continuity testing is usually performed by detecting the presence of on-chip protection circuits. These circuits protect each input and output of the device from the electrostatic discharge (ESD) and other excessive voltage conditions. The ESD protection circuits prevent the input and output pins from exceeding a small voltage above or below the power supply voltage or ground. Diodes and silicon-controlled rectifiers (SCRs) can be used to short the excess currents from the protected pin to ground or to a power terminal.

An ESD protection diode conducts the excess ESD current to ground or power any time, the pin's voltage exceeds one diode drop above (or below) the power or ground voltage. SCRs are similar to ESD protection diodes, but they are triggered by a separate detection circuit. Any of the variety of detection circuits can be used to trigger the SCR when the protected pin's voltage exceeds a safe voltage range.

Once triggered, an SCR behaves like a forward-biased diode from the protected pin to power or ground. The SCR remains in its triggered state until the excessive voltage is removed. Since an SCR behaves much like a diode when triggered, the term "protection diode" is used to describe ESD protection circuits whether they employ a simple diode or a more elaborate SCR structure.

DUT pins may be configured with either one or two protection diodes, connected as shown in the below figure. Notice that the diodes are reverse-biased when the device is powered up assuming normal input and output voltage levels. This effectively makes them "invisible" to the DUT circuits during normal operation.

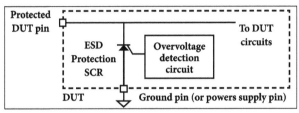

SCR-based ESD protection circuit.

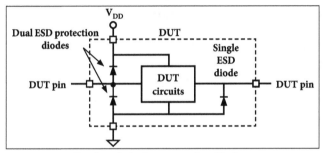

Dual and single protection diodes.

To verify that each pin can be connected to the tester without electrical shorts or open circuits, ATE tester forces a small current across each protection diode in the forward-biased direction. The DUT's power supply pins are set to zero volts to disable all the on-chip circuits and to connect the far end of each diode to ground. ESD protection diodes connected to the positive supply are tested by forcing a current ICONT into the pin as shown in the below figure and measuring the voltage, VCONT, that appears at the pin with respect to ground.

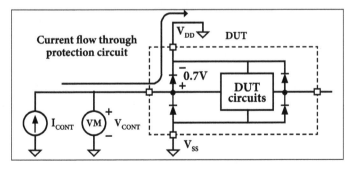

Checking the continuity of the diode connected to the positive supply. The other diode is tested by reversing the direction of the forced current.

If the tester does not see the expected diode drops on each pin, then the continuity test fails and the device is not tested further. Protection diodes connected to the negative supply or ground are tested by reversing the direction of the forced current.

In the case of an SCR-based protection circuit, the current source initially sees an open circuit. Because the current source output tries to force current into an open circuit, its output voltage rises rapidly. The rising voltage soon triggers the SCR's detection circuit. Once triggered, the SCR accepts the current from the current source and the voltage returns to one diode drop above ground. Thus, the difference between a diode-based ESD protection circuit and an SCR-based circuit is hardly noticeable during a continuity test.

The amount of current chosen is typically between 100μA and 1mA, but the ideal value depends on the characteristics of the protection diodes. Too much current may damage the diodes, while too little current may not fully bias them. The voltage drop across a good protection diode usually measures between 550 and 750 mV. For the purpose of illustration, we shall assume that a conducting diode has voltage drop of 0.7 V.

A dead short to ground will result in a reading of 0V, while an open circuit will cause the tester's current source to reach a programmed clamp voltage. Many mixed-signal devices have multiple power supply and ground pins. Continuity to these power and ground connections may or may not be testable. If all supply pins or all ground pins are not properly connected to ground, then continuity to some or all of the non-supply pins will fail.

However, if only some of the supply or ground pins are not grounded, the others will provide a continuity path to zero volts. Therefore, the unconnected power supply or ground pins may not be detected. One way to test the power and ground pins individually is to connect them to ground at a time, using relays to break the connections to the other power and ground pins. Continuity to the power or ground pin can then be verified by looking for the protection diode between it and another DUT pin.

Occasionally, a device pin may not include any protection diodes at all. Continuity to these unprotected pins must be verified by an alternative method, perhaps by detecting a small amount of current leaking into the pin or by detecting the presence of on-chip component such as a capacitor or resistor. Since unprotected pins are highly vulnerable to ESD damage, they are used only in special cases.

One such example is a high-frequency input requiring very low parasitic capacitance. The space-charge layer in the reverse-biased protection diode might add several picofarads of parasitic capacitance to a device pin. Since even a small amount of stray capacitance presents a low impedance to very high-frequency signals, the protection diode must be omitted to enhance the electrical performance of the DUT.

Continuity can be tested one pin at a time, an approach known as serial continuity testing. Unfortunately, serial testing is a time-consuming and costly approach. Modern ATE testers are capable of measuring continuity on all or most pins in parallel rather than measuring the protection diode drops one at a time. These testers accomplish parallel testing using so-called per-pin measurement instruments as shown in figure (a).

Clearly, it is more economical to test all the pins at once using many current sources and voltage meters. Unfortunately, there are a few potential problems to consider. First, a fully parallel test of pins may not detect pin-to-pin shorts. If two device pins are shorted together for some reason, the net current through each diode does not change. Twice as much current is forced through the parallel combination of two diodes. The shorted circuit configuration will therefore result in the expected voltage drop across each diode, resulting in both pins passing the continuity test.

Obviously, the problem can be solved by performing a continuity test on each pin in a serial manner at the cost of extra test time. However, a more economical approach is to test every other pin for continuity on one test pass while grounding the remaining pins. Then the remaining pins can be tested during a second pass while the previously tested pins are grounded. Shorts between adjacent pins would be detected using this dual-pass approach, as illustrated in figure (b).

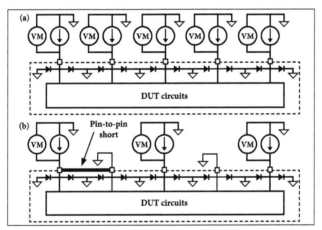

Parallel continuity testing: (a) Full parallel testing with possible adjacent fault masking; (b) Minimizing potential adjacent fault masking by exciting every second pin.

A second, subtler problem with parallel continuity testing is related to the analog measurement performance. Both the analog pins and digital pins must be tested for continuity. On some testers the per-pin continuity test circuitry is limited to digital pins only. The analog pins of the tester may not include per-pin continuity measurement capability. On these testers, continuity testing on the analog pins can be performed one pin at a time using a single current source and voltmeter. These two instruments can be connected to each device pin one at a time to measure protection diode drops.

This is a very time-consuming serial test method, which should be avoided if possible. Alternatively, the analog pins can be connected to the per-pin measurement electronics of digital pins. This allows completely parallel testing of continuity. Unfortunately, the digital per-pin electronics may inject noise into sensitive analog signals. Also, the signal trace connecting the DUT to the per-pin continuity electronics adds a complex capacitive and inductive load to the analog pin, which may be unacceptable.

The signal trace can also behave as a parasitic radio antenna into which unwanted signals can couple into analog inputs. Clearly, full parallel testing of analog pins should be treated with care. One solution to the noise and parasitic loading problems is to isolate each analog pin from its per-pin continuity circuit using a relay. This complicates the DIB design but gives high performance with minimal test time. A tester having per-pin continuity measurement circuits on both the analog and digital pins represents a superior solution.

3.3.1 Leakage Currents

Purpose of Leakage Testing

Each input pin and output pin of a DUT exhibits a phenomenon called leakage. When a voltage is applied to a high-impedance analog or digital input pin, a small amount of current will typically leak into or out of the pin. This current is called leakage current, or simply leakage. Leakage can also be measured on output pins that are placed into a non-driving high-impedance mode. A good design and manufacturing process should result in very low leakage currents. Typically the leak-age is less than 1µA, although this can vary from one device design to another.

One of the main reasons to measure leakage is to detect the improperly processed integrated circuits. Leakage can be caused by many physical defects such as metal filaments and particulate matter that forms shorts and leakage paths between layers in the IC. Another reason to measure leakage is that excessive leakage currents can cause improper operation of the customer's end application. Leakage currents can cause DC offsets and other parametric shifts.

The third reason to test leakage is that excessive leakage currents can indicate a poorly processed device that initially appears to be functional but which eventually fails after a few days or weeks in the customer's product. This type of early failure is known as infant mortality.

Leakage Test Technique

Leakage is measured by simply forcing a DC voltage on the input or output pin of the device under test and measuring the small current flowing into or out of the pin. Unless otherwise specified in the data sheet, leakage is typically measured twice. It is measured once with an input voltage near the positive power supply voltage and again with the input near ground (or negative supply). These two currents are referred as IIH (input current, logic high) and IIL (input current, logic low), respectively.

Digital inputs are typically tested at the valid input threshold voltages, VIH and VIL. Analog input leakage is typically tested at specific voltage levels listed in the data sheet. If no particular input voltage is specified, then the leakage specification applies to the entire allowable input voltage range. Since leakage is usually highest at one or both input voltage extremes, it is often measured at the maximum and minimum allowable input voltages. Output leakage (IOZ) is measured in a manner similar to input leakage, although the output pin must be placed into a high-impedance (HIZ) state using a test mode or other control mechanism.

Serial versus Parallel Leakage Testing

Leakage, like continuity can be tested one pin at a time or all pins at once. Since leakage currents can flow from one pin to another, serial testing is superior to parallel testing from a defect detection perspective. However from a test time perspective, parallel testing is desired. As in continuity testing, a compromise can be achieved by testing every other pin in a dual-pass approach.

Continuity tests are usually implemented by forcing DC current and measuring voltage. Leakage tests are implemented by forcing DC voltage and measuring current. Since the tests are similar in nature, tester vendors generally design both the capabilities into the per-pin measurement circuits of the ATE tester's pin cards.

Analog leakage, like analog continuity is often measured using the per-pin resources of digital pin cards. Again, a tester with per-pin continuity measurement circuits on both the analog and digital pins represents a superior solution, assuming that the extra per-pin circuits are not prohibitively expensive.

Measurement of Offset Currents

The input offset current is the difference between the two input currents driven from a common source.

$$|I_{OS}| = I_a^+ + I_a^-$$

It tells us how much larger one current is than the other. Bias current compensation would work if both bias currents I_B^+ and I_B^- are equal. So, smaller the input offset current better the OP amp. The 741 op-amps have input offset current of 20nA.

3.3.2 Power Supply Currents

Supply Current

Once the continuity test has passed, the next stage is to power up the device in some meaningful manner and check the supply current drawn. This is an important test as it serves to protect both the ATE tester, device under test (DUT) and the socket from damage caused by a bad or misaligned device.

Power supply current trips may also be employed to the same end. Further measurements of the supply current may be made with greater accuracy, and under specified functional conditions, such as device reset, or the device operating at known speeds. In general, the supply current is measured at the maximum supply voltage, as this is usually the condition of maximum supply current draw.

Test Techniques

Most ATE testers are able to measure the current flowing from each voltage source connected to the DUT. Supply currents are therefore very easy to measure in most cases. The power supply is simply set to the desired voltage and the current from its output is measured using one of the tester's ammeters.

When measuring supply currents, the only difficulties arise out of ambiguities in the data sheet.

For example,

Are the analog outputs loaded or unloaded during the supply current test?

Is digital block XYZ operating in mode A, mode B, or idle mode?

In general, it is safe to assume that the supply currents are to be tested under worst-case conditions.

The test engineer should work with the design engineers to attempt to specify the test conditions that are likely to result in the worst-case test conditions. These test conditions should be spelled out clearly in the test plan so that everyone understands the exact conditions used during production testing. Often the actual worst-case conditions are not known until the device has been thoroughly characterized. In these cases, the test program and test plan have to be updated to reflect the characterized worst-case conditions.

Supply currents are often specified under several test conditions, such as power-down mode, standby mode and normal operational mode. In addition, the digital supply currents are specified separately from the analog supply currents. IDD (CMOS) and ICC (bipolar) are commonly used designations for supply current. IDDA, IDDD, ICCA and ICCD are the terms used when analog and digital supplies are measured separately.

Many devices have multiple power supply pins that are connected to a common power supply in normal operation. Design engineers often need to know how much current is flowing into each individual power supply pin. Sometimes, the test engineer can accommodate this requirement by connecting each power supply pin to its own supply. Other times, there are too many DUT supply pins to provide each with its own separate power supply. In these cases, relays can be used to temporarily connect a dedicated power supply to the pin under test.

Another problem that can plague the power supply current tests is settling time. The supply current flowing into a DUT must settle to a stable value before it can be measured. The tester and DIB circuits must also settle to a stable value. This normally takes 5 to 10 ms in normal modes of DUT operation. However, in power-down modes, the specified supply current is often less than 100 μA. Since the DIB usually includes bypass capacitors for the DUT, each capacitor must be allowed to charge until the average current into or out of the capacitor is stable.

The charging process can take hundreds of milliseconds if the current must stabilize within microamps. Some types of bypass capacitors may even exhibit leakage current greater than the current to be measured. A typical solution to this problem is to connect only a small bypass capacitor (say 0.1 µF) directly to the DUT and then connect a larger capacitor (say 10 µF) through a relay as shown in the below figure. The large bypass capacitor can be disconnected temporarily while the power-down current is measured.

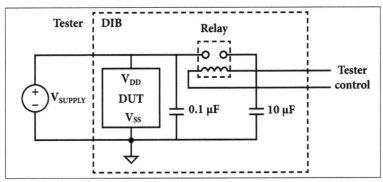

Arranging different-sized bypass capacitors to minimize power supply current settling behaviour.

Input Bias Current

The Op-amp's input is a differential amplifier, which may be made of BJT or FET. In either case, the input transistor must be biased into their linear region by supplying currents into the bases by the external circuit.

In ideal Op-amp, no current is drawn from the input terminals. But practically, op-amp input terminals do conduct a small value of AC current to bias input transistors. Manufacturers specify the input bias current IB as the average value of base currents entering into the terminals of an op-amp as in the below figure.

Bias current is the input current. The transistor (internal transistor of the input differential stage) needs to produce an amount of output voltage.

So, $I_B = \dfrac{I_B^+ + I_B^-}{2}$ ...(1)

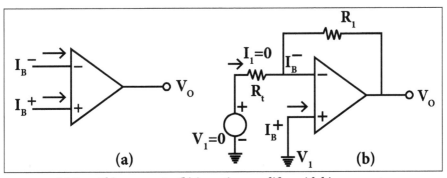

(a) Input bias currents, (b) inverting amplifier with bias currents.

For 741, a bipolar op-amp, the bias current is 500 nA or less. Consider the basic inverting amplifier. If the input voltage V_i is set to zero volts, V_O should be zero volts.

Instead, we find that the output voltage is offset by,

$$V_O = (I_B)R_f \qquad \qquad \qquad ...(2)$$

For Example,

If $R_f = 1M\Omega$, $I_B = 500$ nA, then $V_O = 500$ mV with zero input. Output voltage is 500 mV. This is unacceptable. This effect can be compensated as in figure by adding a resistor R_{comp} between the non-inverting input terminal and ground.

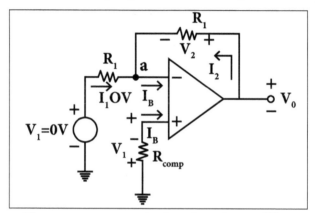

Bias current compensation in an inverting amplifier.

Current I_B^+ flowing through R_{comp} develops a voltage V_1 across it.

Then by KVL, we get,

$$-V_1 + 0 + V_2 - V_0 = 0$$

$$V_0 = V_2 - V_1 \qquad \qquad \qquad ...(3)$$

By selecting proper value of R_{comp} V_2 can be cancelled with V_1 and output V_O will be zero.

$\therefore$ The value of R_{comp} is derived as,

$$V_1 = I_B + R_{comp}$$

$$I_B^+ = \frac{V_1}{R_{comp}} \qquad \qquad \qquad ...(4)$$

The node 'a' is at voltage $(-V_1)$, because the voltage at the non-inverting input terminal is $(-V_1)$. So with $V_i = 0V$, we get,

$$I_1 = \frac{V_i - (-V_t)}{R_t} = 0 + \frac{V_1}{R_1} = \frac{V_1}{R_1} \quad [\because V_i = 0]$$

$$I_1 = \frac{V_1}{R_1} \quad \quad \quad \quad ...(5)$$

Also,

$$I_2 = \frac{V_2}{R_f} \quad \quad \quad \quad ...(6)$$

For compensation, VO should be zero for $V_i = 0V$, i.e. $V_1 = V_2$

$$\therefore \quad I_2 = \frac{V_1}{R_f} \quad \quad \quad \quad ...(7)$$

KCL at node 'a' gives,

$$\Rightarrow I_1 + I_2 = I_B^-$$

Substituting for I_1 and I_2,

$$I_B^- = \frac{V_1}{R_1} = \frac{V_1}{R_f}$$

$$\therefore I_B^- = V_1 \frac{(R_1 + R_f)}{R_1 R_f} \quad \quad \quad \quad ...(8)$$

Assuming $I_B^- = I_B^+$ and using equation (4) and (8) we get,

$$I_B^+ = \frac{V_1}{R_{comp}}$$

$$\therefore \quad \frac{V_1(R_1 + R_f)}{R_1 R_f} = \frac{V_1}{R_{comp}} \quad \quad \quad \quad ...(9)$$

$$R_{comp} = R_1 \| R_f = \frac{R_1 R_f}{R_1 + R_f}$$

To compensate bias currents, R_{comp} should be equal to parallel combination of resistors R_1 and R_f. The effect of input bias current in an amplifier can be compensated by placing the compensating resister R_{comp}.

Measurement of Bias Current

The op-amp's input is a differential amplifier, which may be made of BJT or FET. In either case, the input transistors should be biased into this linear region by supplying currents into the bases.

In an ideal op-amp, no current is drawn from the input terminals. But practically, input terminals conduct a small value of dc current to bias the input transistors when the base currents flow through external resistances, they produce a small differential input voltage or unbalance. This represents a false input signal. When amplified, this small input unbalance produces an offset in the output voltage.

$$I_B = \frac{I_B^+ = I_B^-}{2}$$

For 741, the bias current is 500nA or less. The smaller the input bias current, the smaller the offset at the output voltage.

3.3.3 DC References and Regulators

Voltage Regulators

A voltage regulator is one of the most basic analog circuits. The function of a voltage regulator is to provide a well-specified and constant output voltage level from a poorly specified and sometimes fluctuating input voltage. The output of the voltage regulator would then be used as the supply voltage for other circuits in the system. The below figure illustrates the conversion of a 6 to 12-V ranging power supply to a fixed 5-V output level.

Voltage regulators can be tested using a fairly small number of DC tests. Some of the important parameters for a regulator are output no-load voltage, output voltage or load regulation, input or line regulation, input or ripple rejection and dropout voltage.

Output no-load voltage is measured by simply connecting a voltmeter to the regulator output with no load current and measuring the output voltage V_O.

Load regulation measures the ability of the regulator to maintain the specified output voltage V_O under different load current conditions I_L . As the output voltage changes with increasing load current, one defines the output voltage regulation as the percentage change in the output voltage (relative to the ideal output voltage, $V_{O.NOM}$) for a specified change in the load current.

Load regulation is measured under minimum input voltage conditions as,

$$\text{load regulation} \equiv 100\% \times \frac{\Delta V_o}{V_{o.NOM}} \bigg|_{\max\{\Delta I_L\}\ \minimum V_I} \qquad ...(1)$$

The largest load current change, max (ΔI_L) is created by varying the load current from the minimum rated load current (typically 0mA) to the maximum rated load current.

Load regulation is sometimes specified as the absolute change in voltage, ΔV_0 rather than as a percentage change in V_0. The test definition will be obvious from the specification units (i.e., volts or percentage).

Line regulation or input regulation measures the ability of the regulator to maintain a steady output voltage over a range of input voltages. Line regulation is specified as the percentage change in the output voltage as the input line voltage changes over its largest allowable range. Like the load regulation test, line regulation is sometimes specified as an absolute voltage change rather than a percentage. Line regulation is measured under maximum load conditions as,

$$\text{line regulation} \equiv 100\% \times \left. \frac{\Delta V_0}{V_{0.\text{NOM}}} \right|_{\max\{\Delta I_I\}\text{minimum} I_L} \qquad \text{...(2)}$$

For the regulator shown in the below figure, with the appropriate load connected to the regulator output, the line regulation would be computed by first setting the input voltage to 6 V, measuring the output voltage, then readjusting the input voltage to 12 V and again measuring the output voltage to calculate ΔV_0. The line regulation would then be computed by using (2).

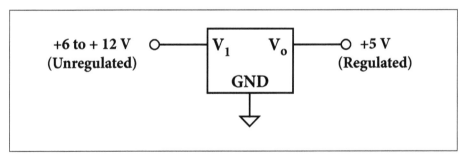

5-V DC voltage regulator.

Input rejection or ripple rejection is the ratio of the maximum input voltage variation to the output voltage swing, measured at a particular frequency (commonly 120 Hz) or a range of frequencies. It is a measure of the circuit's ability to reject periodic fluctuations of rectified AC volt-age signals applied to the input of the regulator. Input rejection can also be measured at DC using the input voltage range and output voltage swing measured during the line regulation test.

Dropout voltage is the lowest voltage that can be applied between the input and output pins without causing the output to drop below its specified minimum output voltage level. Dropout voltage is tested under maximum current loading conditions. It is possible to search for the exact dropout voltage by adjusting the input voltage until the output reaches its mini-mum acceptable voltage, but this is a time-consuming test method.

In production testing, the input can simply be set to the specified dropout voltage plus the minimum acceptable output voltage. The output is then measured to guarantee that it is equal to or above the minimum acceptable output voltage.

Voltage References

Voltage regulators are commonly used to supply the steady voltage while also supplying a relatively large amount of current. However, many of the DC voltages used in a mixed-signal device do not draw a large amount of current. For example, a 1-V DAC reference does not need to supply 500mA of current. For this reason, low-power voltage references are often incorporated into mixed-signal devices rather than high-power voltage regulators.

The output of on-chip voltage references may or may not be accessible from the external pins of a DUT. It is common for the test engineer to request a set of test modes so that reference volt-ages can be measured during production testing. This allows the test program to evaluate the quality of the DC references even if they have no explicit specifications in the data sheet.

The design and test engineers can then determine whether failures in the more complicated AC tests may be due to a simple DC voltage error in the reference circuits. DC reference test modes also allow the test program to trim the internal DC references for more precise device operation.

Ic79xx Fixed Voltage Regulator

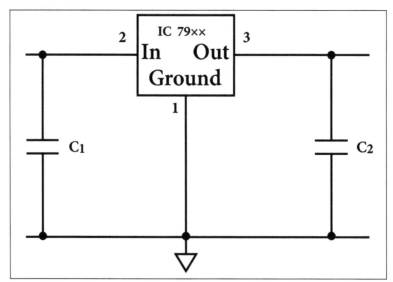

IC79xx Connection diagram.

The pin out configuration of IC 79xx is shown in the figure. The pin 1 acts as the Ground terminal (0V). The pin 2 acts as the input terminal (5V to 24V). The pin 3 acts as the output terminal (constant 5V).

Connection Diagram

IC 79xx is used in circuits in order to improve stability, two capacitors and C_1 and C_2 are used. The capacitor C_1 is used only if the regulator is separated from filter capacitor by more than 3". It must be a 2.2 µF solid tantalum capacitor or 25µF aluminum electrolytic capacitor. The capacitor C_2 is required for stability. Usually 1 µF solid tantalum capacitor is used. We can also use 25µF aluminum electrolytic capacitor. Values given may be increased without limit.

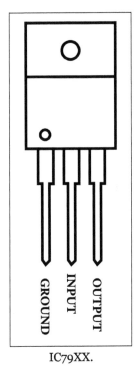

IC79XX.

3.3.4 Impedance Measurements

Input Impedance

Input impedance (Z_{IN}), also referred to as input resistance, is a common specification for analog inputs. In general, impedance refers to the behavior of both resistive and reactive components in the circuit. Hence, impedance and resistance refer to the same quantity at DC.

Input impedance is a simple measurement to make. If the input voltage is a linear function of the input current (i.e., if it behaves according to Ohm's law), then one simply forces a voltage V and measures a current I or vice versa, and computes the input impedance according to,

$$Z_{IN} = \frac{V}{I} \qquad \qquad ...(1)$$

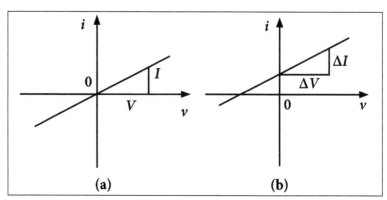

Input i-v characteristic curves resulting in an impedance function with (a) equal DC and AC operation and (b) unequal DC and AC operation.

Figure (a) illustrates the input i-v relationship of a device satisfying Ohm's law. Here we see that the i-v characteristic is a straight line passing through the origin with a slope equal to 1/ZIN.

In many instances, the i-v characteristic of an input pin is a straight line but does not pass through the origin as shown in Figure (b). Such situations typically arise from biasing considerations where the input terminal of a device is biased by a constant current source such as that shown in the below figure or has in series with it an unknown voltage source to ground, or in series between two components comprising the input series impedance.

In cases such as these, we cannot use (1) to compute the input impedance, as it will not lead correctly to the slope of the i-v characteristic. Instead, we can measure the change in the input current (ΔI) that results from a change in the input voltage (ΔV) and computes the input impedance using,

$$Z_{IN} = \frac{\Delta V}{\Delta I} \qquad\qquad \ldots(2)$$

If the input impedance is so low that it would cause excessive currents to flow into the pin, another approach is needed. The alternative method is to force two controlled currents and measure the resulting voltage difference. This is often referred to as a force-current/measure-voltage method. Input impedance is again calculated using (2).

Large changes in voltages and currents are easier to measure than small ones. The test engineer should beware of saturating the input of the device with excessive voltages, though. Saturation could lead to extra input current resulting in an inaccurate impedance measurement. The device data sheet should list the acceptable range of input voltages.

Output Impedance

Output impedance (Z_{OUT}) is measured in the same way as input impedance. It is typically much lower than the input impedance. So, it is usually measured using a force-current/

measure-voltage technique. However, in cases where the output impedance is very high, it may be measured using the force-voltage/measure-current method instead.

Differential Impedance Measurements

Differential impedance is measured by forcing two differential voltages and measuring the differential current change or by forcing two differential currents and measuring the differential voltage change.

3.3.5 DC Offset Measurements

V_{MID} and Analog Ground

Many analog and mixed-signal integrated circuits are designed to operate on a single power supply voltage (V_{DD} and ground) rather than a more familiar bipolar supply (V_{DD}, V_{SS} and ground). Often these single-supply circuits generate their own low-impedance voltage between V_{DD} and ground that serves as a reference voltage for the analog circuits. This reference voltage, which we will refer as V_{MID}, may be placed halfway between V_{DD} and ground or it may be placed at some other fixed voltage such as 1.35 V.

In some cases, V_{MID} may be generated off-chip and supplied as an input voltage to the DUT. To simplify the task of circuit analysis, we can define any circuit node to be 0 V and measure all other voltages relative to this node. Therefore, in a single-supply circuit having a V_{DD} of 3 V, V_{SS} connected to ground and an internally generated V_{MID} of 1.5 V, we can redefine all the voltages relative to the V_{MID} node.

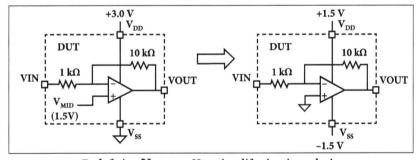

Redefining V_{MID} as 0 V to simplify circuit analysis.

Using this definition of 0 V, we can translate our single-supply circuit into a bipolar configuration with V_{DD} = +1.5 V, V_{MID} = 0 V, and V_{SS} = -1.5 V. For this approach to be valid, it is assumed that no hidden impedance lies between V_{SS} and ground, which is a reasonable assumption at low to moderate frequencies, less at very high frequencies.

Analog ground is a term used in the test and measurement industry to refer to a high-quality ground that is separated from the noisy ground connected to the DUT's digital circuits. In fact, the term "ground" has a definite meaning when working with measurement equipment since it is actually tied to earth ground for safety reasons.

Here, we will use the term analog ground to refer to a quiet oV voltage for use by analog circuits and use the term V_{MID} to refer to an analog reference voltage (typically generated on-chip) that serves as the IC's analog "ground."

DC Transfer Characteristics (Gain and Offset)

The input-output DC transfer characteristic for an ideal amplifier is shown in the below figure. The input-output variables of interest are voltage, but they could just as easily be replaced by cur-rent signals. As the real world is rarely accommodating to IC and system design engineers, the actual transfer characteristic for the amplifier would deviate somewhat from the ideal or expected curve. To illustrate the point, we superimpose another curve on the plot and label it as "Typical."

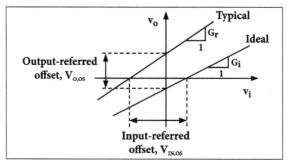

Amplifier input–output transfer characteristics in its linear region.

In order to maintain correct system operation, design engineers require some assurance that the amplifier transfer characteristic is within acceptable tolerance limits. The particular interests to the test engineer are the gain and offset voltages shown in the figure.

Input Offset Voltage

Input offset voltage is the voltage applied at the input terminals of an op-amp to make output zero, typically of the order of millivolts. Whenever both the input terminals of the op-amp are ground ideally, the output voltage should be zero. However, the practical op-amp shows a small non-zero output voltage.

To make this output voltage zero, one may have to apply a small voltage at the input terminals. This voltage is called input offset voltage V_{os}, since positive terminal (+) voltage is around $V_1 = oV$.

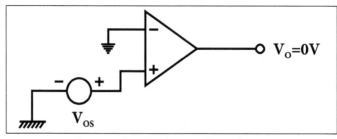

Op-amp showing input offset voltage.

The voltage V_2 at (-) input terminal is given by,

$$V_2 = \left(\frac{R_1}{R_1 + R_f}\right)V_0$$

$$V_0 = \left(\frac{R_1 + R_f}{R_1}\right)V_2$$

$$V_0 = \left(1 + \frac{R_f}{R_1}\right)V_2$$

Since, $V_{iOS} = |V_1 - V_2|$

And, $V_1 = 0V$

$$V_{OS} = |0 - V_2| = V_2$$

$$\therefore \quad V_0 = \left(1 + \frac{R_f}{R_1}\right)V_{iOS}$$

Thus, the output offset voltage of an op-amp in closed loop configuration (inverting and non inverting) is obtained.

Input Offset Current

Input Offset Current is the algebraic difference between the currents into the inverting and non-inverting terminals.

Input Offset Current, $I_{io} = |I_{b1} - I_{b2}|$

Where,

I_{b1} - Non-inverting input current

I_{b2} - Inverting input current

The maximum input offset current value for 741 IC is 200nA. This value decreases as the matching between the two input terminals is improved and may reduce down to almost 6nA.

The difference in magnitudes of I_B^+ and I_B^- is called as input offset current (I_{os}). Bias current compensation will work if both I_B^+ and I_B^- are equal. Since the input transistors are not identical, there is a small difference between I_B^+ and I_B^-. This difference is called as input offset current (I_{os}) and can be written as,

$$|I_{OS}| = I_B^+ - I_B^-$$

The absolute value of sign indicates that there is no way to predict which of the bias currents will be larger.

Offset current I_{os} for BJT op-amp is 200 nA. Offset current I_{os} for FET is 10 pA. Even with bias current compensation, I_{os} will produce a output voltage when $V_i = 0V$.

$$V_1 = I_B^+ R_{comp}$$

and,

$$I_1 = \frac{V_1}{R_1} = \frac{I_B^+ R_{comp}}{R_1}$$

KCL at node 'a' $\Rightarrow I_1 + I_2 = I_B^-$

$$I_2 = I_B^- - I_1$$

$$I_2 = I_B^- - \left(\frac{I_B^- R_{comp}}{R_1} \right)$$

$$V_o = V_2 - V_1$$
$$V_o = I_2 R_f - V_1$$
$$= I_2 R_f - I_B^+ R_{comp}$$
$$= \left(I_B^- - \frac{I_B^+ R_{comp}}{R_1} \right) R_f - I_B^+ R_{comp}$$

Substituting for $R_{comp} = R_1 \| R_f$

$$V_o = I_B^- R_f - I_B^+ R_{comp} \left[\frac{R_f}{R_1} + 1 \right]$$

$$= I_B^+ R_f - I_B^+ \frac{R_1 R_f}{R_1 + R_f} \left[\frac{R_1 + R_f}{R_1} \right]$$

$$V_o = R_f \cdot \left[I_B^+ - I_B^- \right]$$

$$V_o = R_f \cdot I_{OS}$$

So even with current bias compensation and with feedback resistor of 1MΩ, a 741 BJT Op-amp has an output offset voltage,

$$V_o = 1M\Omega \times 200nA$$

= 200 mV, with zero input voltage.

It can be seen that effect of offset current can be minimized by having a small value of R_f. Unfortunately, R_1 must be kept large to have high input impedance. With R_1 large, R_f also should be kept large so as to have reasonable gain. This will allow large feedback resistance, while keeping the resistance to ground (seen by the inverting input) low as shown in dotted in the network.

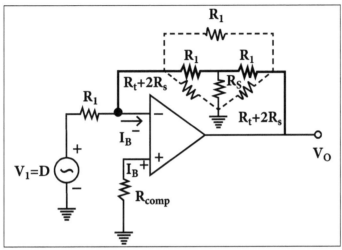

Inverting amplifier with 'T' feedback network.

The 'T' Network provides a feedback signal as if the network were a single feedback resistor. By T to Π conversion,

$$R_f = \frac{R_t^2 + 2R_t R_s}{R_3}$$

$$R_t R_3 = R_t^2 + 2R_t R_s$$

$$R_t^2 = R_t R_s + 2R_t R_s$$

$$R_t^2 = R_s(R_f R_t)$$

To design a T-network,

$$R_t \ll \frac{R_f}{2}$$

Then calculate $R_s = \dfrac{R_t^2}{R_f - 2R_t}$

Measurement of Offset Voltage

Ideally, the output voltage must be zero when the voltage between the inverting and non-inverting inputs is zero. In reality, the output voltage might not be zero with zero input voltage. This is due to unavoidable imbalances, mismatches, tolerances and so on inside the op-amp.

In order to make output voltage zero, we have to apply a small voltage at the input terminals to make output voltage zero. This voltage is called input offset voltage. i.e., input offset voltage is the voltage required to be applied at input for making output voltage to zero volts.

The 741 op-amp has input offset voltage of 5mV under no signal conditions. Therefore, we may have to apply a differential input of 5mV, to produce an output voltage of exactly zero.

3.3.6 DC Gain Measurements

Since the OP-AMP amplifies the difference voltage between the two input terminals, the voltage gain of the amplifier is defined as,

$$\text{Voltage gain} = \frac{\text{Output voltage}}{\text{Differential input voltage}}$$

$$A = \frac{V_o}{V_{id}}$$

Because output signal amplitude is much large than the input signal the voltage gain is commonly called as large signal voltage gain. For 741C is voltage gain is 200,000 typically.

Closed-Loop Gain

Closed-loop DC gain is one of the simplest measurements to make, because the input-output signals are roughly comparable in level. Closed-loop gain, denoted as G, is defined as the slope of the amplifier input-output transfer characteristic, as illustrated in the below figure. We refer this gain as closed-loop because it typically contrived from a set of electronic devices configured in a negative feedback loop.

It is computed by simply dividing the change in output level of the amplifier or circuit by the change in its input.

$$G = \frac{\Delta V_o}{\Delta V_1} \qquad\qquad ...(1)$$

DC gain is measured using two DC input levels that fall inside the linear region of the amplifier. This latter point is particularly important, because false gain values are often obtained when the amplifier is unknowingly driven into saturation by poorly chosen input levels. The range of linear operation should be included in the test plan.

Gain can also be expressed in decibels (dB). The conversion from volt-per-volt to decibels is simply expressed as,

$$G\big|_{dB} = 20 \log_{10}|G| \qquad\qquad ...(2)$$

The logarithm function in (2) is a base-10 log as opposed to a natural log. Gain may also be specified for circuits with differential inputs and/or outputs. The measurement is basically the same.

Differential measurements can be made by measuring each of the two output voltages individually and then computing the difference mathematically. Alternatively, a differential voltmeter can be used to directly measure differential voltages. Obviously, the differential voltmeter approach will work faster than making two separate measurements. Therefore, the use of a differential voltmeter is the preferred technique in production test programs.

Sometimes, the differential voltage is very small compared to the DC offset of the two DUT outputs. A differential voltmeter can often give more accurate readings in these cases.

In cases requiring extreme accuracy, it may be necessary to measure the input voltages as well as the output voltages. The DC voltage sources in most ATE testers are well calibrated and stable enough to provide a voltage error no greater than 1mV in most cases.

If this level of error is unacceptable, then it may be necessary to use the tester's high-accuracy voltmeter to measure the exact input voltage levels rather than trusting the sources to produce the desired values. Moreover, they do not depend on any value for the offsets, only that the appropriate slope is obtained from the linear region of the transfer characteristic.

Open-Loop Gain

Open-loop gain (abbreviated G_{ol}) is a basic parameter of op amps. It is defined as the gain of the amplifier with no feedback path from output to input. Since many op amps have G_{ol} values of 10,000 V/V or more, it is difficult to measure open-loop gain with the straightforward techniques. It is difficult to apply a voltage directly to the input of an open-loop op amp without causing it to saturate, forcing the output to one power supply rail or the other.

For example, if the maximum output level from an op amp is ±5 V and its open-loop gain is equal to 10,000 V/V, then an input-referred offset of only 500 µV will cause the amplifier output to saturate. Since many op amps have input-referred offsets ranging over several millivolts, we cannot predict the input voltage range that results in the unsaturated output levels.

We can overcome this problem using a second op amp connected in a feedback path as shown in the below figure. The second amplifier is termed as a nulling amplifier. The nulling amplifier forces its differential input voltage to zero through a negative feedback loop formed by resistor string R2 and R1, together with the DUT op amp. This loop is also termed as a servo loop.

By doing so, the output of the op amp under test can be forced to a desired output level according to,

$$V_{O,DUT} = 2V_{MID} - V_{SRC1} \qquad \qquad ...(3)$$

Where,

V_{MID} is a DC reference point and V_{SRC1} is the programmed DC voltage from SRC1.

The nulling amplifier and its feedback loop compensate for the input-referred offset of the DUT amplifier. This ensures that the DUT output does not saturate due to its own input-referred offset.

The two matched resistors, R_3 are normally chosen to be around 100 kΩ as a compromise between the source loading and op-amp bias-induced offsets. Since the gain around the loop is extremely large, feedback capacitor C is necessary to stabilize the loop. The capacitance value of 1 to 10nF is usually sufficient. R_{LOAD} provides the specified load resistance for the G_{ol} test.

Under steady-state conditions, the signal that is fed back to the input of the DUT amplifier denoted as $V_{IN,DUT}$ is directly related to the nulling amplifier output V_{O-NULL} according to,

$$V_{IN.DUT} = V_{IN.DUT}^{+} - V_{IN.DUT}^{-} = \frac{R_1}{R_1 + R_2}\left(V_{o.NULL} - V_{MID}\right) \qquad ...(4)$$

Where $V_{IN.DUT}^{+}$ and $V_{IN.DUT}^{-}$ are the positive and negative inputs to the DUT amplifier, respectively.

Subsequently, the open-loop voltage gain of the DUT amplifier is found from (1), (3) and (4) as,

$$G_{ol} = \frac{\Delta V_{O.DUT}}{\Delta V_{IN.DUT}} = -\left(\frac{R_1 + R_2}{R_1}\right)\frac{\Delta V_{SRC1}}{\Delta V_{O.NULL}} \qquad ...(5)$$

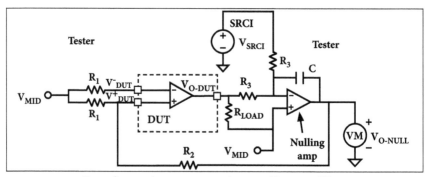

Open-loop gain test setup using a nulling amplifier.

The nulling loop method allows the test engineer to force two desired outputs and then indirectly measure the tiny inputs that caused those two outputs. In this manner, very large gains can be measured without measuring tiny voltages. The accuracy of this approach depends on accurately knowing the values of R_1 and R_2, as well as on selecting the two resistors labeled as R_3 in the above figure.

In order to maximize the signal handling capability of the test setup, as well as avoid saturating the nulling amplifier, it is a good idea to set the voltage divider ratio to a value approximately equal to the inverse of the expected open-loop gain of the DUT op amp.

$$\frac{R_1}{R_1 + R_2} = \frac{1}{G_{Ol}} \qquad \qquad ...(6)$$

From which we can write $R_2 \approx G_{ol} R_1$.

A detailed circuit analysis reveals that this offset is caused exclusively by the input-referred offset of the DUT. Hence, the offset that appears at the output of the nulling amplifier, denoted as $V_{O,NULL,OS}$, can be used to compute the input-referred offset of the DUT, $V_{IN,DUT,OS}$.

Input-referred offset would be calculated using,

$$V_{IN,DUT,OS} = \frac{R_1}{R_1 + R_2} V_{O,NULL,OS} \qquad \qquad ...(7)$$

Because this method involves the same measured data used to compute the open-loop gain, it is a commonly used method to determine the op amp input-referred offset.

3.3.7 DC Power Supply Rejection Ratio

The Power Supply Rejection Ration (PSRR) is the ratio of the variation of the output voltage due to a change in the power supply voltage. This is almost always specified as a DC characteristic of the OP-Amp, not an AC characteristic.

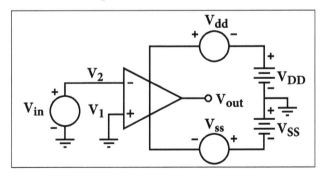

PSRR.

$$PSRR = \frac{A_v (V_{dd} = 0)}{A_{dd} (V_{in} = 0)}$$

3.3.8 DC Common-Mode Rejection Ratio

CMRR

- In OP-amp, the output voltage is proportional to the difference between the voltages applied to its two input terminals.

- When two input voltages are equal, ideally, the output voltages should be zero.

- A signal applied to both input terminals of the OP-amp is called as common-mode signal. Usually it is an unwanted noise signal.

- The ability of an op-amp to suppress the common-mode signals is expressed in terms of its common-mode rejection ratio (CMRR).

- Typically the CMRR for a 741 op-amp is around 90 dB.

CMRR=20 $\log_{10}$ (Differential Voltage Gain/Common Mode Gain) dB.

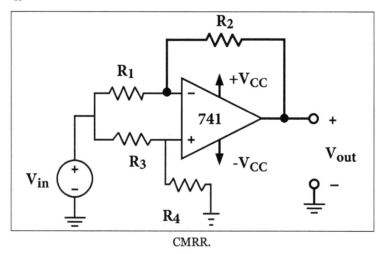

CMRR.

3.3.9 Comparator DC Tests

Input Offset Voltage

Input offset voltage for a comparator is defined as the differential input voltage that causes the comparator to switch from one output logic state to the other. The differential input voltage can be ramped from one voltage to another to find the point at which the comparator changes state. This switching point is, however, dependent on the input common-mode level.

We usually tests for the input offset voltage under worst-case conditions as outlined in the device test plan.

Threshold Voltage

Sometimes, a fixed reference voltage is supplied to one input of the comparator, forming a circuit known as a slicer. The input offset voltage specification is typically replaced by a single-ended specification, known as threshold voltage.

The slicer in the below figure is tested in a manner similar to that of the comparator circuit. Assuming that the threshold voltage is expected to fall between 1.45 and 1.55 V, the input voltage from SRC1 is ramped upward from 1.45 to 1.55 V. The output switches states when the input is equal to the slicer's threshold voltage.

Notice that the threshold voltage will be affected by the accuracy of the on-chip voltage reference, VTH. The threshold voltage should be equal to the sum of the slicer's reference voltage VTH plus the input offset voltage of the comparator. Threshold voltage error is defined as the difference between the actual and ideal threshold voltages.

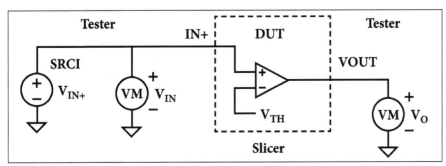

Slicer threshold voltage test setup.

Hysteresis

In the comparator input offset voltage, the output changed when the input voltage reached 5mV. This occurred on the rising input voltage. On the falling input voltage, the threshold may change to a lower voltage. This characteristic is termed as hysteresis, and it may or may not be an intentional design feature.

Hysteresis is defined as the difference in threshold voltage between a rising input test condition and a falling input condition. It should be noted that the input offset voltage and hysteresis may change with different common-mode input voltages. Worst-case test conditions should be determined during the characterization process.

3.3.10 Voltage Search Techniques

Binary Searches Versus Step Searches

The technique of ramping input voltage levels until an output condition is met is termed as a ramp search or step search. Step searches are time-consuming and not well suited for production testing. Instead, binary or linear search methods are often used.

To gain a better understanding of these methods, let us consider the general search process. In mathematical terms, let us denote the input-output behavior of some device under test with some mathematical function, say $y = f(x)$, where x is the input and y is the output. Subsequently, to establish the output at some arbitrary level, say $y = D$, we need to find the value of x that satisfies $f(x)-D = 0$.

If the inverse of is known, then we can immediately solve for input as $x = f^{-1}(D)$. Generally, $f(x)$ is not known, since it is specific to each and every device under test. However, through a source-measurement process, the behavior of $y = f(x)$ is encapsulated in the form of a look-up table or by the direct action of the measurement.

Consequently, through some search process, we can identify the value of x that satisfies $f(x)-D = 0$. As x is a root of the equation $f(x)-D$, the procedures used to identify the root are known as root-finding algorithms. There are numerous root-finding algorithms such as bisection (binary), secant (linear), false-position and Newton-Raphson methods. Any one of these can be adapted to test.

Binary Search Method

To determine the input value such that the output equals a desired target value within some tolerance, start with two input values. One value is selected such that the output is greater than some desired target value, while the other is selected to obtain an output less than this target value.

Let us denote these two output conditions as output @ input1 and output @ input2. Subsequently, the binary search process can be described using the following pseudo code:

```
TARGET = desired value

Measure output @ input1

Measure output @ input2

output @ input3 =1000 # initialization

Do WHILE | output @ input3 - TARGET | ≥ tolerance value

        Set input3 = (input1 + input2)/2

        Measure output @ input3

        IF output @ input3 - TARGET is of opposite sign to output @ input1
- TARGET Do:

                Set input1 = input1 and input2 = input3

        ELSE

                Set input1 = input3 and input2 = input2

        ENDIF

END Do
```

A binary search can be applied to the comparator input offset voltage test. Instead of ramping the input voltage from 1.45 to 1.55 V, the comparator input is set half-way between to 1.5 V and the output is observed. If the output is high, then the input is increased by one-quarter of the 100-mV search range (25 mV) to try to make the output go low.

If the output is low, then the input is reduced by 25 mV to try to force the output high. Then the output is observed again. This time, the input is adjusted by one-eighth of the search range (12.5mV). This process is repeated until the desired input adjustment resolution is reached.

The problem with the binary search technique is that it does not work well in the presence of hysteresis. The binary search algorithm assumes that the input offset voltage is the same whether the input voltage is increased or decreased. If the comparator exhibits hysteresis, then there are two different threshold voltages to be measured. To get around this problem without reverting to the time-consuming ramp search technique, a hybrid approach can be used.

A binary search can be used to find the approximate threshold voltage quickly. Then a step search can be used with a much smaller search voltage range. Another solution to the hysteresis problem is to use a modified binary search algorithm in which the output state of the comparator is returned to a known logic state between binary search approximations. This can be achieved by forcing the input either well above or well below the threshold voltage. In this way, steps are always taken in one direction, avoiding hysteresis effects.

To measure hysteresis, a binary search is used once with the output state forced high between approximations. Then the input offset is measured again with the output state forced low between approximations. The difference in input offset readings is equal to the hysteresis of the comparator.

Linear Searches

Linear circuits can make use of an even faster search technique known as a linear search. A linear search is similar to the binary search, except that the input approximations are based on a linear interpolation of input-output relationships. For example, if a 0mV input to a buffer amplifier results in a 10mV output and a 1mV input results in a 20mV output, then a -1mV input will probably result in a 0mV output. The linear search algorithm keeps refining its guesses using a simple straight-line approximation $V_{OUT} = M \times V_{IN} + B$ algorithm until the desired accuracy is reached.

We can easily generalize the principles described above to one involving the following set of iteration equations with pseudo code:

```
TARGET = desired value

k=1 # initialize iteration

Measure output @ inputk-1
```

```
Measure output @ inputk

output @ inputk+1 = 1000 # initialization

Do WHILE | output @ inputk+1 - TARGET | ≥ tolerance value
```

$$\text{Set input}_{k+1} = \text{input}_k - \left(\text{output}@\text{input}_k - \text{TARGET}\right)\times$$

$$\left(\frac{\text{input}_k - \text{input}_{k-1}}{\text{output}@\text{input}_k - \text{output}@\text{input}_{k-1}}\right)$$

```
Measure output @ inputk+1

k = k + 1

END Do
```

3.3.11 DC Tests for Digital Circuits

$$I_{IH} / I_{IL}$$

The data sheet for a mixed-signal device usually lists several DC specifications for digital inputs and outputs. Input leakage is also specified for digital output pins that can be set to a high-impedance state.

$$V_{IH}/V_{IL}$$

The input high voltage (V_{IH}) and input low voltage (V_{IL}) specify the threshold voltage for digital inputs. It is possible to search for these voltages using a binary search or step search, but it is more common to simply set the tester to force these levels into the device as a go/no-go test.

If the device does not have adequate V_{IH} and V_{IL} thresholds, then the test program will fail one of the digital pattern tests that are used to verify the DUT's digital functionality. To allow a distinction between pattern failures caused by V_{IH}/V_{IL} settings and patterns failing for other reasons, the test engineer may add a second identical pattern test that uses more forgiving levels for V_{IH} / V_{IL}.

If the digital pattern test fails with the specified V_{IH} / V_{IL} levels and passes with the less demanding settings, then VIH/VIL thresholds are the likely failure mode.

$$V_{OH}/V_{OL}$$

V_{OH} and V_{OL} are the output equivalent of V_{IH} and V_{IL}. V_{OH} is the minimum guaranteed voltage for an output when it is in the high state. V_{OL} is the maximum guaranteed voltage when the output is in the low state. These voltages are usually tested in two ways.

First, they are measured at DC with the output pin set to static high/low levels. Sometimes a pin cannot be set to a static output level due to poor design for test considerations, so only a dynamic test can be performed.

Dynamic V_{OH}/V_{OL} testing is performed by setting the tester to expect high voltages above V_{OH} and low voltages below V_{OL}. The tester's digital electronics are able to verify these voltage levels as the outputs toggle during the digital pattern tests. Dynamic V_{OH}/V_{OL} testing is another go/no-go test approach, since the actual VOH/VOL voltages are verified but not measured.

I_{OH}/I_{OL}

V_{OH} and V_{OL} levels are guaranteed while the outputs are loaded with specified load currents, I_{OH} and I_{OH}. The tester must pull current out of the DUT pin when the output is high. This load current is called I_{OH}.

Likewise, the tester forces the I_{OL} current into the pin when the pin is low. These currents are intended to force the digital outputs closer to their V_{OH}/V_{OL} specifications, making the V_{OH}/V_{OL} tests more difficult for the DUT to pass. I_{OH} and I_{OL} are forced using a diode bridge circuit in the tester's digital pin card electronics.

I_{OSH} and I_{OSL} Short-Circuit Current

Digital outputs often include a current-limiting feature that protects the output pins from damage during short-circuit conditions. If the output pin is shorted directly to ground or to a power supply pin, the protection circuits limit the amount of current flowing into or out of the pin.

Short-circuit current is measured by setting the output to a low state and forcing a high voltage (usually V_{DD}) into the pin. The current flowing into the pin (I_{OSL}) is measured with one of the tester's current meters. Then the output is set to a high state and 0V is forced at the pin. The current flowing out of the pin (I_{OSH}) is again measured with a current meter.

3.4   Measurement Accuracy

The cause of escapes in an analog or mixed-signal environment is largely related to the measurement process itself. A value presented to the instrument by the DUT will introduce errors, largely on account of the electronic circuits that make up the instrument. These errors manifest themselves in various forms.

1. Accuracy and Precision

- Accuracy: The difference between the average of measurements and a standard sample for which the "true" value is known. The degree of conformance of a test instrument to absolute standards, usually expressed as a percentage of reading or a percentage of measurement range (full scale).

- Precision: The variation of a measurement system obtained by repeating the measurements on the same sample back-to-back using the same measurement conditions.

According to these definitions, precision refers only to the repeatability of a series of measurements. It does not refer to consistent errors in the measurements. A series of measurements can be incorrect by 2V, but as long as they are consistently wrong by the same amount, then the measurements are considered to be precise.

Many sources of error can affect the accuracy of a given measurement. The accuracy of a measurement should probably refers to all the possible sources of error. However, the accuracy of an instrument is often specified in the absence of repeatability fluctuations and instrument resolution limitations.

2. Systematic or Bias Errors

Systematic or bias errors are those that show up consistently from measurement to measurement. For example, assume that an amplifier's output exhibits an offset of 100 mV from the ideal value of 0V. Using a digital voltmeter (DVM), we could take multiple readings of the offset over time and record each measurement. A typical measurement series might look like this:

101 mV, 103 mV, 102 mV, 101 mV, 102 mV, 103 mV, 103 mV, 101 mV, 102 mV . . .

This measurement series shows an average error of about 2mV from the true value of 100mV. Errors like this are caused by consistent errors in the measurement instruments. The errors can result from a combination of many things, including DC offsets, gain errors, and non-ideal linearity in the DVM's measurement circuits. Systematic errors can often be reduced through a process called calibration.

3. Random Errors

In the preceding example, notice that the measurements are not repeatable. The DVM gives readings from 101 to 103mV. Such variations do not surprise most engineers because DVMs are relatively inexpensive.

Inexperienced test engineers are sometimes surprised to learn that an expensive tester cannot give perfectly repeatable answers. They may be inclined to believe that the tester software is defective when it fails to produce the same result every time the program is executed. However, the experienced test engineers recognize that a certain amount of random error is to be expected in the analog and mixed-signal measurements.

Random errors are usually caused by thermal noise or other noise sources in the DUT or the tester hardware. One of the biggest challenges in mixed-signal testing is determining whether the random errors are caused by bad DIB design, by bad DUT design, or by the tester itself. If the source of error is found and cannot be corrected by a design change, then averaging or filtering of measurements may be required.

4. Resolution (Quantization Error)

In the 100-mV measurement list, notice that the measurements are always rounded off to the nearest millivolt. The measurement may have been rounded off by the person taking the measurements, or perhaps the DVM was only capable of displaying three digits.

ATE measurement instruments have similar limitations in the measurement resolution. Limited resolution results from the fact that continuous analog signals must first be converted into a digital format before the ATE computer can evaluate the test results. The tester converts the analog signals into digital form using analog-to-digital converters (ADCs).

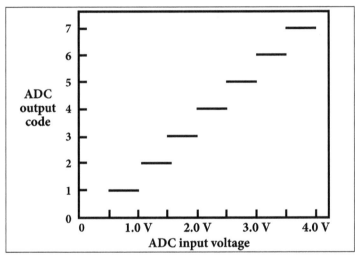

Output codes versus input voltages for an ideal 3-bit ADC.

ADCs by nature exhibit a feature called quantization error. Quantization error is a result of the conversion from an infinitely variable input voltage to a finite set of possible digital output results from the ADC. Figure below shows the relationship between the input voltages and output codes for an ideal 3-bit ADC. Notice that an input voltage of 1.2 V results in the same ADC output code as an input voltage of 1.3 V. In fact, any voltage from 1.0 to 1.5 V will produce an output code of 2.

If this ADC were the part of a crude DC voltmeter, the meter would produce an output reading of 1.25 V any time the input voltage falls between 1.0 and 1.5 V. This inherent error in ADCs and measurement instruments is caused by quantization error. The resolution of a DC meter is often limited by the quantization error of its ADC circuits.

If a meter has 12 bits of resolution, it means that it can resolve a voltage to one part in $2^{12}-1$(one part in 4095). If the meter's full-scale range is set to ±2 V, then a resolution of approximately 1mV can be achieved (4 V/4095 levels). This does not automatically mean that the meter is accurate to 1mV, it simply means the meter cannot resolve variations in input voltage smaller than 1mV.

An instrument's resolution can far exceed its accuracy. For example, a 23-bit voltmeter might be able to produce a measurement with a 1-μV resolution, but it may have a systematic error of 2 mV.

5. Repeatability

Non-repeatable answers are a fact of life for mixed-signal test engineers. A large portion of the time required to debug a mixed-signal test program can be spent tracking down the various sources of poor repeatability. Since all electrical circuits generate a certain amount of random noise, measurements such as those in the 100-mV offset example are fairly common.

In fact, if a test engineer gets the same answer 10 times in a row, it is time to start looking for a problem. Most likely, the tester instrument's full-scale voltage range has been set too high, resulting in a measurement resolution problem. For example, if we configured a meter to a range having a 10-mV resolution, then our measurements from the prior example would be very repeatable (100 mV, 100 mV, 100 mV, 100 mV, etc.).

A novice test engineer might think that this is a terrific result, but the meter is just rounding off the answer to the nearest 10-mV increment due to an input ranging problem. Unfortunately, a voltage of 104 mV would also have resulted in this same series of perfectly repeatable, perfectly incorrect measurement results. Repeatability is desirable, but it does not guarantee accuracy.

6. Stability

A measurement instrument's performance may drift with time, temperature and humidity. The degree to which a series of supposedly identical measurements remains constant over time, temperature, humidity and all other time-varying factors is referred as stability. Stability is an essential requirement for accurate instrumentation.

Shifts in the electrical performance of measurement circuits can lead to errors in the tested results. Most shifts in performance are caused by temperature variations. Testers are usually equipped with the temperature sensors that can automatically determine when a temperature shift has occurred. The tester must be recalibrated anytime the ambient temperature has shifted by a few degrees.

The calibration process brings the tester instruments back into alignment with known electrical standards so that measurement accuracy can be maintained at all times. After the tester is powered up, the tester's circuits must be allowed to stabilize to a constant temperature before calibrations can occur. Otherwise, the measurements will drift over time as the tester heats up.

When the tester chassis is opened for maintenance or when the test head is opened up or powered down for an extended period, the temperature of the measurement electronics will typically drop.

Calibrations then have to be rerun once the tester recovers to a stable temperature.

Shifts in performance can also be caused by aging electrical components. These changes are typically much slower than shifts due to temperature. The same calibration processes used to account for temperature shifts can easily accommodate shifts of components caused by aging. Shifts caused by humidity are less common, but can also be compensated for by periodic calibrations.

7. Correlation

Correlation is another activity that consumes a great deal of mixed-signal test program debug time. Correlation is the ability to get the same answer using different pieces of hardware or software. It can be extremely frustrating to try to get the same answer on two different pieces of equipment using two different test programs. It can be even more frustrating when two supposedly identical pieces of test equipment running the same program give two different answers.

In general, it is a good idea to make sure that the correlation errors are less than one-tenth of the full range between the minimum test limit and the maximum test limit. However, this is just a rule of thumb. The exact requirements will differ from one test to the next. Whatever correlation errors exist, they must be considered part of the measurement uncertainty, along with non-repeatability and systematic errors.

The test engineer must consider several categories of correlation. Test results from a mixed-signal test program cannot be fully trusted until the various types of correlation have been verified. The more common types of correlation include tester-to-bench, tester-to-tester, program-to-program, DIB-to-DIB, and day-to-day correlation.

i. Tester-to-Bench Correlation

Often, a customer will construct a test fixture using bench instruments to evaluate the quality of the device under test. Bench equipment such as oscilloscopes and spectrum analyzers can help validate the accuracy of the ATE tester's measurements. Bench correlation is a good idea, since ATE testers and test programs often produce incorrect results in the early stages of debug.

In addition, IC design engineers often build their own evaluation test setups to allow quick debug of device problems. Each of these test setups must correlate to the answers given by the ATE tester. Often the tester is correct and the bench is not. Other times, test program problems are uncovered when the ATE results do not agree with a bench setup. The test engineer will often need to help debug the bench setup to get to the bottom of correlation errors between the tester and the bench.

ii. Tester-to-Tester Correlation

Sometimes a test program will work on one tester, but not on another presumably identical tester. The differences between testers may be catastrophically different, or they may be very subtle. The test engineer should compare all the test results on one tester to the test results obtained using other testers. Only after all the testers agree on all tests is the test program and test hardware debugged and ready for production.

Similar correlation problems arise when an existing test program is ported from one tester type to another. Often, the testers are neither software compatible nor hardware compatible with one another. In fact, the two testers may not even be manufactured by the same ATE vendor.

A myriad of correlation problems can arise because of the vast differences in DIB layout and tester software between different tester types. To some extent, the architecture of each tester will determine the best test methodology for a particular measurement.

The given test may have to be executed in a very different manner on one tester versus another. Any difference in the way a measurement is taken can affect the results. For this reason, correlation between two different test approaches can be very difficult to achieve. Conversion of a test program from one type of tester to another can be one of the most daunting tasks a mixed-signal test engineer faces.

iii. Program-to-Program Correlation

When a test program is streamlined to reduce test time, the faster program must be correlated to the original program to make sure no significant shifts in measurement results have occurred. Often, the test reduction techniques cause measurement errors because of reduced DUT settling time and other timing-related issues. These correlation errors must be resolved before the faster program can be released into production.

iv. DIB-to-DIB Correlation

No two DIBs are identical and sometimes the differences cause correlation errors. The test engineer should always check to make sure that the answers obtained on multiple DIB boards agree. DIB correlation errors can often be corrected by focused calibration software written by the test engineer.

v. Day-to-Day Correlation

Correlation of the same DIB and tester over a period of time is also important. If the tester and DIB have been properly calibrated, there should be no drift in the answers from one day to the next. Subtle errors in software and hardware often remain hidden until day-to-day correlation is performed. The usual solution to this type of correlation problem is to improve the focused calibration process.

8. Reproducibility

The difference between reproducibility and repeatability relates to the effects of correlation and stability on a series of supposedly identical measurements. Repeatability is most often used to describe the ability of a single tester and DIB board to get the same answer multiple times as the test program is repetitively executed.

Reproducibility is the ability to achieve the same measurement result on a given DUT using any combination of equipment and personnel at any given time. It is defined as the statistical deviation of a series of supposedly identical measurements taken over a period of time. These measurements are taken using various combinations of test conditions that ideally should not change the measurement result.

For example, the choice of equipment operator, tester, DIB board and so on, should not affect any measurement result.

Consider the case in which a measurement is highly repeatable, but not reproducible. In such a case, the test program may consistently pass a particular DUT on a given day and yet consistently fail the same DUT on another day or on another tester. Clearly, measurements must be both the repeatable and reproducible to be production-worthy.

3.5   Calibration and Checkers: Dealing with Measurement Errors

Measurement accuracy can be improved by eliminating the bias error β associated with the measurement process. The following are the several ways to remove this error so that the tester is performing with maximum accuracy at all times during its operation.

1. Traceability to Standards

Every tester and bench instrument must ultimately correlate to standards maintained by a central authority such as the National Institute of Standards and Technology (NIST). In the United States, this government agency is responsible for maintaining the standards for pounds, gallons, inches and electrical units such as volts, amperes and ohms. The chain of correlation between the NIST and the tester's measurements involves a series of calibration steps that transfers the "golden" standards of the NIST to the tester's measurement instruments.

Many testers have a centralized standards reference, which is a thermally stabilized instrument in the tester mainframe. The standards reference is periodically replaced by a freshly calibrated reference source. The old source is sent back to the certified calibration laboratory, which recalibrates the reference so that it agrees with the NIST standards.

Similarly, bench instruments are periodically recalibrated so that they too are traceable to the NIST standards. By periodically refreshing the tester's traceability link to the NIST, all the testers and bench instruments can be made to agree with one another.

2. Hardware Calibration

Hardware calibration is a process of physical "knob tweaking" that brings a piece of measurement instrumentation back into agreement with calibration standards. For instance, oscilloscope probes often include a small screw that can be used to nullify the overshoot in rapidly rising digital edges. This is one of the common examples of hardware calibration.

One major problem with hardware calibration is that it is not a convenient process. It generally requires a manual adjustment of a screw or knob. Robotic screwdrivers might be employed to allow partial automation of the hardware calibration process. However, the use of robotics is an elaborate solution to the calibration problem. Full automation can be achieved using a simpler procedure known as software calibration.

- Modeling a voltmeter with the ideal voltmeter and a nonideal component in cascade.

- Calibrating the nonideal effects using a software routine.

3. Software Calibration

Using software calibration, ATE testers are able to correct the hardware errors without adjusting any physical knobs. The basic idea behind the software calibration is to separate the instrument's ideal operation from its non-idealities. Then a model of the instrument's non ideal operation can be constructed, followed by the correction of the non-ideal behavior using a mathematical routine written in software. The below figure illustrates this idea for a voltmeter.

In figure (a), a "real" voltmeter is modeled as a cascade of two parts:

- An ideal voltmeter.

- A black box that relates the voltage across its input terminals v_{DUT} to the voltage that is measured by the ideal voltmeter, $v_{measured}$.

This relationship can be expressed in more mathematical terms as,

$$v_{measured} = f(v_{DUT}) \quad\quad ...(1)$$

Where f(·) indicates the functional relationship between $v_{measured}$ and v_{DUT}.

The true functional behavior f(·) is seldom known. Thus, we may assumes a particular behavior or model, such as a first-order model given by,

$$v_{measured} = Gv_{DUT} + offset \quad\quad ...(2)$$

Where G and offset are the gain and offset of the voltmeter, respectively.

These values must be determined from the measured data. Subsequently, a mathematical procedure is written in software that performs the inverse mathematical operation,

$$v_{calibrated} = f^{-1}v_{measured} \quad\quad ...(3)$$

Where $v_{\text{calibrated}}$ replaces v_{DUT} as an estimate of the true voltage that appears across the terminals of the voltmeter as depicted in figure (b). If $f(\cdot)$ is known precisely, then $v_{\text{calibrated}} = v_{\text{DUT}}$.

In order to establish an accurate model of an instrument, precise reference levels are necessary. The number of reference levels required to characterize the model fully will depend on its order. i.e., the number of parameters used to describe the model. For the linear or first order model, it has two parameters, G and offset. Hence, two reference levels will be required.

To avoid conflict with the meter's normal operation, relays are used to switch in these reference levels during the calibration phase. For example, the voltmeter in the below figure includes a pair of calibration relays, which can connect the input to two separate reference levels, V_{ref1} and V_{ref2}.

During a system level calibration, the tester closes one relay and connects the voltmeter to Vref1 and measures the voltage, which we shall denote as $v_{\text{measured1}}$. Subsequently, this process is repeated for the second reference level Vref2 and the voltmeter provides a second reading, $v_{\text{measured2}}$.

Based on the assumed linear model for the voltmeter, we can write two equations in terms of two unknowns,

$$v_{\text{measured1}} = GV_{\text{ref1}} + \text{offset} \quad\quad ...(4)$$

$$v_{\text{measured2}} = GV_{\text{ref2}} + \text{offset} \qu\quad ...(5)$$

Using linear algebra, the two model parameters can then be solved to be,

$$G = \frac{v_{\text{measured2}} - v_{\text{measured1}}}{V_{\text{ref2}} - V_{\text{ref1}}} \qu\quad ...(6)$$

And

$$\text{offset} = \frac{v_{\text{measured1}} V_{\text{ref2}} - v_{\text{measured2}} V_{\text{ref1}}}{V_{\text{ref2}} - V_{\text{ref1}}} \qu\quad ...(7)$$

The parameters of the model, G and offset, are also known as calibration factors or cal factors for short.

When subsequent DC measurements are performed, they are corrected using the stored calibration factors according to,

$$v_{\text{calibrated}} = \frac{v_{\text{measured}} - \text{offset}}{G} \qu\quad ...(8)$$

This expression is found by isolating v_{DUT} on one side of the expression in (2) and replacing it by $v_{calibrated}$.

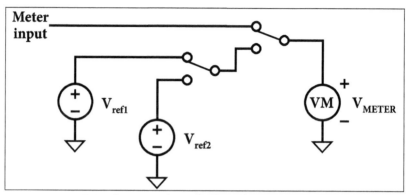

DC voltmeter gain and offset calibration paths.

Most testers use much more elaborate calibration schemes to account for linearity errors and other nonideal behavior in the meter's ADC and associated circuits. Also, the meter's input stage can be configured many ways, and each of these possible configurations needs a separate set of calibration factors.

For example, if the input stage has 10 different input ranges, then each range setting requires a separate set of calibration factors. The calibration factors are measured and stored automatically during the tester's periodic system calibration and checker process.

4. System Calibrations and Checkers

Testers are calibrated on a regular basis to maintain traceability of each instrument to the tester's calibration reference source. In addition to calibrations, software is also executed to verify the functionality of hardware and make sure it is production worthy. This software is called a checker program or checker for short.

Often calibrations and checkers are executed in the same program. If a checker fails, the repair and maintenance (R&M) staff replaces the failing tester module with a good one. After replacement, the new module must be completely recalibrated.

There are several types of calibrations and checkers. These include calibration reference source replacement, performance verification (PV), periodic system calibrations and checkers, instrument calibrations at load time and focused calibrations. A common replacement cycle time for calibration sources is once every six months.

To verify that the tester is in compliance with all its published specifications, a more extensive process called performance verification may be performed. Although full performance verification is typically performed at the tester vendor's production floor, it is seldom performed on the production floor. By contrast, periodic system calibrations and checkers are performed on a regular basis in a production environment. These software calibration and checker programs verify that all the system hardware is production worthy.

Since tester instrumentation may drift slightly between system calibrations, the tester may also perform a series of fine-tuning calibrations each time a new test program is loaded. The extra calibrations can be limited to the subset of instruments used in a particular test program. This helps to minimize the program load time.

To maintain accuracy throughout the day, these calibrations may be repeated on a periodic basis after the program has been loaded. They may also be executed automatically if the tester temperature drifts by more than a few degrees. Finally, focused calibrations are often required to achieve maximum accuracy and to compensate for non-idealities of DIB board components such as buffer amplifiers and filters.

Unlike the ATE tester's built-in system calibrations, focused calibration and checker software is the responsibility of the test engineer. Focused calibrations fall into two categories:

- Focused instrument calibrations.

- Focused DIB calibrations and checkers.

i. Focused Instrument Calibrations

Testers typically contain a combination of slow, accurate instruments and fast instruments that may be less accurate. The accuracy of the faster instruments can be improved by periodically referencing them back to the slower more accurate instruments through a process called focused calibration. Focused calibration is not always necessary. However, it may be required if the test engineer needs higher accuracy than the instrument is able to provide using the built-in calibrations of the tester's operating system.

A simple example of focused instrument calibration is a DC source calibration. The DC sources in a tester are generally quite accurate, but occasionally they need to be set with minimal DC level error.

A calibration routine that determines the error in a DC source's output level can be added to the first run of the test program. A high-accuracy DC voltmeter can be used to measure the actual output of the DC source. If the source is in error by 1 mV, for instance, then the requested voltage is reduced by 1 mV and the output is retested. It may take several iterations to achieve the desired value with an acceptable level of accuracy.

A similar approach can be extended to the generation of a sinusoidal signal requiring an accurate RMS value from an arbitrary waveform generator (AWG). A high-accuracy AC voltmeter is used to measure the RMS value from the AWG. The discrepancy between the measured value and the desired value is then used to adjust the programmed AWG signal level. The AWG output level will thus converge toward the desired RMS level as each iteration is executed. Another application of focused instrument calibration is spectral leveling of the output of an AWG.

An important application of AWGs is to provide a composite signal consisting of N sine waves or tones all having equal amplitude at various frequencies and arbitrary phase. Such waveforms are in a class of signals commonly referred to as multitone signals. Mathematically, a multitone signal y(t) can be written as,

$$y(t)A_o + A_1 \sin(2\pi f_1 t + \phi_1) + \cdots + A_N \sin(2\pi f_N t + \phi_N) = A_o + \sum_{k=1}^{N} A_k \sin(2\pi f_k t + \phi_k)$$

where A_k, f_k and ϕ_k denotes the amplitude, frequency and phase respectively of the kth tone.

A multitone signal can be viewed in either the time domain or in the frequency domain. Time domain views are analogous to oscilloscope traces, while frequency-domain views are analogous to spectrum analyzer plots. The frequency-domain graph of a multitone signal contains a series of vertical lines corresponding to each tone frequency and whose length * represents the root-mean-square (RMS) amplitude of the corresponding tone. Each line is referred to as a spectral line.

Figure below illustrates the time and frequency plots of a composite signal consisting of three tons of frequencies 1, 2.5 and 4.1 kHz, all having an RMS amplitude of 2 V.

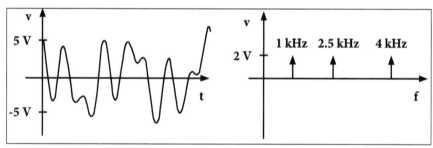

Time-domain and frequency-domain views of a three-tone multitone.

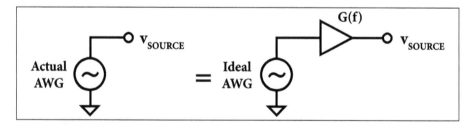

Modeling an AWG as a cascaded combination of an ideal source and frequency-dependent gain block.

The peak amplitude of each sinusoid in the multitone is simply $\sqrt{2} \times 2$ or 2.82 V, so we could just easily plot these values as peak amplitudes rather than RMS.

The AWG produces its output signal by passing the output of a DAC through a low-pass anti-imaging filter. Due to its frequency behavior, the filter will not have a perfectly flat magnitude response. The DAC may also introduce frequency-dependent errors. Thus the amplitudes of the individual tones may be offset from their desired levels. We can therefore model this AWG multitone situation as illustrated in the above figure.

The model consists of an ideal source connected in cascade with a linear block whose gain or magnitude response is described by G(f), where f is the frequency expressed in hertz. To correct for the gain change with frequency, the amplitude of each tone from the AWG is measured individually using a high-accuracy AC voltmeter.

The ratio between the actual output and the requested output corresponds to G(f) at that frequency. This gain can be stored as a calibration factor that can subsequently be retrieved to correct the amplitude error at that frequency. The calibration process is repeated for each tone in the multitone signal.

The composite signal can then be generated with corrected amplitudes by dividing the previous requested amplitude at each frequency by the corresponding AWG gain calibration factor. Because the calibration process equalizes the amplitudes of each tone, the process is called multitone leveling.

As testers continue to evolve and improve, it may become increasingly unnecessary for the test engineer to perform focused calibrations of the tester instruments. Focused calibrations were once necessary on almost all tests in a test program. Today, they can sometimes be omitted with little degradation in accuracy. Nevertheless, the test engineer must evaluate the need for focused calibrations on each test. Even if calibrations become unnecessary in the future, the test engineer should still understand the methodology so that test programs on older equipment can be comprehended.

Calibration of circuits on the DIB, on the other hand, will probably always be required. The tester vendor has no way to predict what kind of buffer amplifiers and other circuits will be placed on the DIB board. The tester operating system will never be able to provide automatic calibration of these circuits. The test engineer is fully responsible for understanding the calibration requirements of all the DIB circuits.

ii. Focused DIB Circuit Calibrations

Often circuits are added to a DIB board to improve the accuracy of a particular test or to buffer the weak output of a device before sending it to the tester electronics. As the signal-conditioning DIB circuitry is added in cascade with the test instrument, a model of the test setup is identical to that given in figure (a).

The only difference is that functional block $v_{measured} = f(v_{DUT})$ includes both the meter and the DIB's behavior. As a result, the focused instrument calibration can be used with no modifications. Conversely, the meter may already have been calibrated so that the functional block $f(\cdot)$ covers the DIB circuitry only. We must keep track of the extent of the calibration to avoid any double counting.

When buffer amplifiers are used to assist the measurement of AC signals, a similar calibration process must be performed on each frequency that is to be measured. Like the AWG calibration, the buffer amplifier also has a non-ideal frequency response and will affect the reading of the meter. Its gain variation, together with the meter's frequency response, must be measured at each frequency used in the test during a calibration run of the test program.

Assuming that the meter has already been calibrated, the frequency response behavior of the DIB circuitry must be correctly accounted for. This can be achieved by measuring the gain in the DIB's signal path at each specific test frequency. Once found, it is stored as a calibration factor. If additional circuits such as filters, ADCs and so on, are added on the DIB board and used under multiple configurations, then each unique signal path must be individually calibrated.

3.5.1 Basic Data Analysis

Data analysis is the process by which we examine the test results and draw conclusions from them. Using data analysis, we can evaluate the DUT design weaknesses, identify DIB and tester repeatability and correlation problems, improve test efficiency and expose test program bugs.

Debugging is one of the main activities associated with the mixed-signal test and product engineering. Debugging activities account for about 20% of the average workweek. Consequently, data analysis plays a very large part in the overall test and product-engineering task.

Many types of data visualization tools have been developed to help us make sense of the reams of test data that are generated by a mixed-signal test program. The ability of the engineer is to model large quantity of data with simple mathematical models. Often, these are based on the assumption of the structure of underlying random behavior and several quantities obtained from the measured data, such as mean and standard deviation.

Data Visualization Tools

Datalogs (Data Lists)

A datalog, or data list, is a concise listing of test results generated by a test program. Datalogs are the primary means by which test engineers evaluate the quality of a tested device. The format of a datalog typically includes a test category, test description, minimum and maximum test limits, and a measured result. The exact format of datalogs varies from one tester type to another, but datalogs all convey similar information.

Lot Summaries

Lot summaries are generated after all devices in a given production lot have been tested. A lot summary lists a variety of information about the production lot, including the lot number, product number, operator number, and so on. It also lists the yield loss and cumulative yield associated with each of the specified test bins. The overall lot yield is defined as the ratio of the total number of good devices divided by the total number of devices tested.

$$\text{lot yield} = \frac{\text{total good devices}}{\text{total devices tested}} \times 100\%$$

Wafer Maps

A wafer map displays the location of failing die on each probed wafer in a production lot. Unlike lot summaries, which only show the number of devices that fail each test category, wafer maps show the physical distribution of each failure category. This information can be very useful in locating areas of the wafer where a particular problem is most prevalent.

3.6   Tester Hardware: Mixed-Signal Tester

General-Purpose Testers Versus Focused Bench Equipment

General-purpose mixed-signal testers must be capable of testing a variety of dissimilar devices. On any given day, the same mixed-signal tester may be expected to test video palettes, cellular telephone devices, data transceivers and general-purpose ADCs and DACs. The test requirements for these various devices are very different from one another.

For example, the cellular telephone base-band interface may require a phase trajectory error test or an error vector magnitude test.

Dedicated bench instruments can be purchased that are specifically designed to measure these application-specific parameters. It would be possible to install one of these stand-alone boxes into the tester and communicate with it through an IEEE-488 GPIB bus. However, if every type of DUT required two or three specialized pieces of bolt-on hardware, the tester would soon resemble Frankenstein's monster and would be prohibitively expensive.

The mixed-signal production tester cannot be focused toward a specific type of device if it is to handle a variety of DUTs. Instead of implementing tests like phase trajectory error and error vector magnitude using a dedicated bench instrument, the tester must emulate this type of equipment using a few general-purpose instruments. The instruments are combined with software routines to simulate the operation of the dedicated bench instruments.

Generic Tester Architecture

Mixed-signal testers come in a variety of "flavors" from a variety of vendors. Unlike the ubiquitous PC, testers are not at all standardized in architecture. Each ATE vendor adds special features that they feel will give them a competitive advantage in the marketplace. Consequently, mixed-signal testers from different vendors do not use a common software platform.

Furthermore, a test routine implemented on one type of tester is often difficult to translate to another tester type because of subtle differences in the hardware tradeoffs designed into each tester. Nevertheless, most mixed-signal testers share many common features.

3.6.1 DC resources

General-Purpose Multimeters

Most testers incorporate a high-accuracy multimeter that is capable of making fast DC measurements. A tester may also provide a slower, very high-accuracy voltmeter for more demanding measurements such as those needed in focused calibrations. However, this slower instrument may not be usable for production tests because of the longer measurement time.

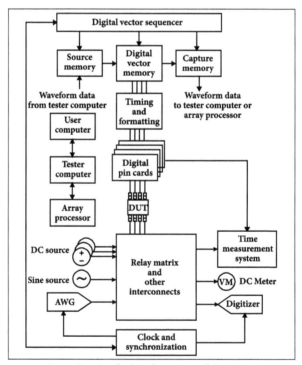

Generic mixed-signal tester architecture.

The fast, general-purpose multimeter is used for most of the production tests requiring a nominal level of accuracy. The above figure shows generic mixed-signal tester architecture. It includes DC sources, system computers, relay control lines, DC meters, time measurement hardware, relay matrix lines, waveform digitizers, arbitrary waveform generators, clocking and synchronization sources, and a digital subsystem for generating and evaluating the digital patterns and signals.

A detailed DC multimeter structure is shown in the below figure. This meter can handle either single-ended or differential inputs. Its architecture includes a high-impedance differential to single-ended converter, a low-pass filter, a programmable gain amplifier (PGA) for input ranging, a high-linearity ADC, integration hardware, and a sample-and-difference stage. It also includes an input multiplexer stage to select one of several input signals for measurement.

The instrumentation amplifier provides a high-impedance differential input. The high impedance avoids potential DC offset errors caused by bias current leaking into the meter. For single-ended measurements, the low end of the meter may be connected to ground through relays in the input selection multiplexer. The multimeter can also be connected to any of the tester's general-purpose DC voltage sources to measure their output voltage.

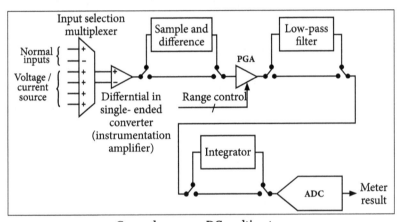

General-purpose DC multimeter.

The meter can also measure current flowing from any of the DC sources. This capability is very useful for measuring the power supply currents, impedances, leakage currents, and other common DC parametric values. A PGA placed before the meter's ADC allows proper ranging of the instrument to minimize the effects of the ADC's quantization error.

The meter may also include a low-pass filter in its input path. The low-pass filter removes high-frequency noise from the signal under test, improving the repeatability of DC measurements. This filter can be enabled or bypassed using software commands. It may also have a programmable cutoff frequency so that the test engineer can make trade-offs between measurement repeatability and test time. In addition, some meters may include an integration stage, which acts as a form of hardware averaging circuit to improve measurement repeatability.

Finally, a sample-and-difference stage is included in the front end of many ATE multimeters. The sample-and-difference stage allows highly accurate measurements of small differences between two large DC voltages.

During the first phase of measurement, a hardware sample-and-hold circuit samples a voltage. This first reference voltage is then subtracted from a second voltage using an amplifier-based subtractor. The difference between the two voltages is then amplified and measured by the meter's ADC, resulting in a high-resolution measurement of the difference voltage. This process reduces the quantization error that would otherwise results from a direct measurement of the large voltages using the meter's higher voltage ranges.

General-Purpose DC Voltage/Current Sources

Most testers include general-purpose DC voltage/current sources, commonly referred to as V/I sources or DC sources. These programmable power supplies are used to provide the DC voltages and currents necessary to power up the DUT and stimulate its DC inputs. Many general-purpose supplies can force either voltage or current, depending on the testing requirements. On most testers, these supplies can be switched to multiple points on the DIB board using the tester's DC matrix.

The system's general-purpose meter can be connected to any DC source to measure its output voltage or its output current. Figure below shows a conceptual block diagram of a DC source having a differential Kelvin connection. A differential Kelvin connection consists of four lines (high force, low force, high sense and low sense) for forcing highly accurate DC voltages.

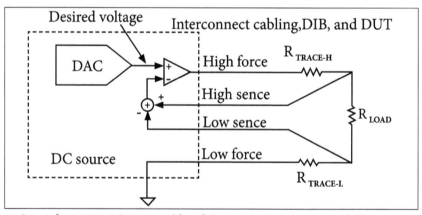

General-purpose DC source with Kelvin connections (conceptual diagram).

The Kelvin connection forms a feedback loop that allows the DC source to force an accurate differential voltage through the resistive wires between the source and DUT. Without the Kelvin connection, the small resistance in the force line interconnections ($R_{TRACE-H}$ and $R_{TRACE-L}$) would cause a small IR voltage drop. The voltage drop would be proportional to the current through the DUT load (R_{LOAD}).

The small IR voltage drop would result in errors in the voltage across the DUT load. The sense lines of a Kelvin connection carry no current. Therefore, they are immune to errors caused by IR voltage drops.

A sense line is provided on the high side of the DC source and also on the low side of the source. The low side sense line counteracts the parasitic resistance in the current return path. Since most instruments are referenced to ground, the low sense lines for all the DC instruments in a tester are often lumped into a single ground sense signal called DZ (device zero), DGS (device ground sense), or some other vendor-specific nomenclature.

This is one of the most important signals in a mixed-signal tester, since it connects the DUT's ground voltage back to the tester's instruments for use as the entire test system's 0-V "golden zero reference". If any voltage errors are introduced into this ground reference signal relative to the DUT's ground, all the instruments will produce DC voltage offsets.

Precision Voltage References

Mixed-signal testers sometimes include high-accuracy, low-noise voltage references. These voltage sources can be used in place of the general-purpose DC sources when the noise and accuracy characteristics of the standard DC source are inadequate. One common example of a precision voltage reference application is the voltage reference for a high-resolution ADC or DAC.

Any noise and DC error on the DC reference of an ADC or DAC translates directly into gain error and increased noise, respectively, in the output of the converter. A precision voltage reference is sometimes used to solve this problem.

Testers may also include non-programmable user power supplies with high output current capability. These fixed supplies provide common power supply voltages (±5 V, ±15 V, etc.) for DIB circuits such as op amps and relay coils. This allows DIB circuits to operate from inexpensive fixed power supplies having high current capability instead of tying up the tester's more expensive programmable DC sources.

Calibration Source

The purpose of a calibration source is to provide traceability of standards back to a central agency such as the National Institute of Standards and Technology (NIST). The calibration source must be recalibrated on a periodic basis (six months is a common period). Often, the source is removed from the tester and sent to a certified standards lab for recalibration.

The old calibration source is replaced by a freshly calibrated one so that the tester can continue to be used in production. On some testers, the high-accuracy multimeter serves as the calibration source. Also, some testers may have multiple instruments that serve as the calibration sources for various parameters such as voltage or frequency.

Relay Matrices

A relay matrix is a bank of electromechanical relays that provides flexible interconnections between many different tester instruments and the DUT. There may be several types of relay matrix in a tester, but they all perform a similar task. At different points in a test program, a particular DUT input may require a DC voltage, an AC waveform, or a connection to a voltmeter.

A relay matrix allows each instrument to be connected to a DUT pin at the appropriate time as illustrated in the below figure. General-purpose relay matrices are used to connect and disconnect various circuit nodes on the DIB board. They have no hardwired connections to tester instruments. Therefore, the purpose and functionality of a general-purpose relay matrix depends on the test engineer's DIB design.

Instrument Relay Matrix.

It allows flexible interconnections between specific tester instruments and pins of the DUT through connections on the DIB board. In addition to relay matrices, many other relays and signal paths are distributed throughout a mixed-signal tester to allow flexibility in interconnections without adding unnecessary relays to the DIB board. The exact architecture of relays, matrices and signal paths varies widely from one ATE vendor's tester to the next.

Relay Control Lines

Despite the high degree of interconnection flexibility provided by the general-purpose relay matrix and other instrument interconnect hardware, there are always cases where a local DIB relay (placed near the DUT) is imperative. Usually, the need for a local DIB relay is driven by the performance of the DUT. For example, there is no better way to get a low-noise ground signal to the input of a DUT than to provide a local relay placed on the DIB directly between the DUT input and the DUT's local ground plane.

Certainly, it is possible to feed the local ground through a DIB trace, through a remote relay matrix, and back through another DIB trace, but this connection scheme invariably leads to poor analog performance. The DIB traces are, after all, radio antennae. Many noise problems can be traced to poor layout of ground connections between the DUT and its ground plane. Local DIB relays minimize the radio antenna effect. Local DIB relays are also used to connect device outputs to various passive loads and other DIB circuits.

The test program controls the local DIB relays, opening and closing them at the appropriate time during each test. The relay coils are driven by the tester's relay control lines. A relay control line driver is shown in the below figure. On some testers, the control line is capable of reading back the state of the voltage on the control line through a readback comparator. The readback comparator allows a low-cost method for determining the state of a digital signal.

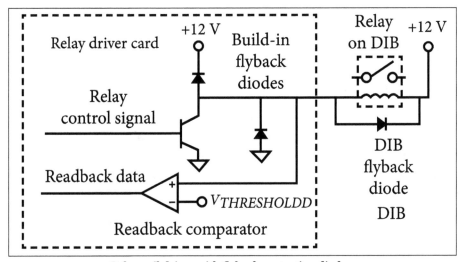

Relay coil driver with flyback protection diodes.

Relay coils produce an inductive kickback when the current is suddenly changed between the on and off states. The inductive kickback, or flyback as it is known, is induced according to the inductance formula $v(t) = L\, di/dt$. Since high kickback voltages could potentially damage the output circuits of the relay driver, its output circuits contain flyback protection diodes to shunt the excess voltage to a DC source or to ground. Many test engineers also add flyback diodes across the coils of the relay.

The extra diode is probably redundant. However, many engineers consider it good practice to add extra fly-back diodes even though they ate up quite a bit of DIB board space. To eliminate the board space issue, the test engineer can choose slightly more expensive relays with built-in flyback diodes.

3.6.2 Digital Subsystem

Digital Vectors

A mixed-signal tester must test the digital circuits as well as mixed-signal and analog circuits. The mixed-signal and digital-only sections of the DUT are exercised using the tester's digital subsystem. The digital subsystem can present high, low, and high-impedance (HIZ) logic levels to the DUT. It can also compare the outputs from the DUT against expected responses to determine whether the digital logic of the DUT has been manufactured without defects.

The tester applies a sequence of drive data to the device and simultaneously compares outputs against expected results. Each drive/compare cycle is called a digital vector. A series of digital vectors is termed as a digital pattern. Vectors of a digital pattern are usually sourced at a constant frequency, although some testers allow the period of each vector to be set independently. The ability to change digital timing on a vector-by-vector basis is commonly called timing on the fly.

Digital Signals

In addition to the simple pass/fail digital pattern tests, the tester must also be capable of sourcing and capturing digital signals. Digital signals are digitized representations of continuous waveforms such as sine waves and multitones.

Digital signals are distinct from digital vectors in that they typically carry analog signal information rather than purely digital information. Usually, the samples of a digital signal must be applied to a DUT along with a repetitive digital pattern that keeps the device active and initiates DAC and/or ADC conversions. Each cycle of the repeating digital pattern is termed as frame.

During a mixed-signal test, the repeating frame vectors must be combined with the non-repeating digital signal sample information to form a repetitive sampling loop. Combining the digital frame vectors with digital signal data, a long sequence of waveform samples can be sent to or captured from the DUT with a very short digital frame pattern. In effect, the sampling frame results in a type of data compression that minimizes the amount of vector memory needed for the tester's digital subsystem.

Looping frames are commonly used when testing DACs and ADCs. A sequence of samples must be loaded into a DAC to produce a continuous sequence of voltages at the DAC's output. In the case of ADC testing, digital signals must be captured and stored into a bank of memory as the looping frame initiates each ADC conversion.

Source Memory

When testing DACs, the digital signal samples representing the desired DAC analog waveform are typically computed in the tester's main test program code. The digital signal samples are stored into a digital subsystem memory block called source memory. The digital frame data, on the other hand, are stored in vector memory. To generate a repeating frame with a new sample for each loop, the contents of the vector memory and source memory are spliced together in real time as the digital pattern is executed.

A digital signal can be modified quickly without changing the frame loop pattern because its data are generated algorithmically by the main test program. The ability to quickly modify the digital signal data is especially useful during the DUT debug and characterization phase.

For example, a DAC may normally be tested using a 1-kHz sine wave digital signal. During the DAC characterization phase, however, the frequency might be swept from 100 Hz to 10 kHz to look for problem areas in the DAC's design. This would be impossibly cumbersome if the digital pattern had to be generated using an expanded, non-looping sequence of ones and zeros.

In fact, some tester architectures attempt to substitute deep, non-looping vector memory in place of source memory. This may reduce the cost of tester hardware, but it invariably results in frustrated users. One of the main differences between a mixed-signal tester and a digital tester with bolt-on analog instruments is the presence of source and capture memories in the digital subsystem.

Capture Memory

Devices such as ADCs produce a series of digitized waveform samples that must be captured and stored into a bank of memory called capture memory (or receive memory). Capture memory serves the opposite function of source memory. Each time the sampling frame is repeated, the digital output from the device is stored into the capture memory. The capture memory address pointer is incremented each time a digital sample is captured. Once a complete set of samples have been collected, they are transferred to an array processor or to the tester computer for analysis.

Pin Card Electronics

The pin card electronics for each digital channel are located inside the test head on most mixed-signal testers. A pin card electronics board may actually contain multiple channels of identical circuitry. Each channel's circuits consist of a programmable driver, a programmable comparator, various relays, dynamic current load circuits, and other circuits necessary to drive and receive signals to and from the DUT.

A generic digital pin card is shown in the below figure. The driver circuitry consists of a fixed impedance driver (typically 50 Ω) with two programmable logic levels, VIH and VIL. These levels are controlled by a pair of driver-level DACs whose voltages are controlled by the test program. The driver can also switch into a high-impedance state (HIZ) at any point in the digital pattern to allow data to come from the DUT into the pin card's comparator.

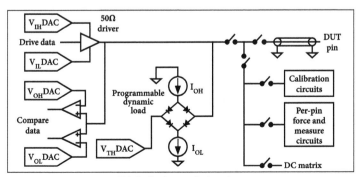

Digital pin card circuits.

The driver circuits may also include programmable rise and fall times, though fixed rise and fall times are more common. Normally, the fixed rise and fall times are designed to be as fast as the ATE vendor can make them. Rise and fall times between 1 ns and 3 ns are typical in today's testers.

The comparator also has two programmable logic levels, V_{OH} and V_{OL}. These are also controlled by another pair of DACs whose voltages are controlled by the test program. The pin card comparator is actually a pair of comparators, one for the VOH level and one for VOL.

If the DUT signal is below VOL, then the signal is considered a logic low. If the DUT is above VOH, then it is considered logic high. If the DUT output is between these thresholds, then the output state is considered a midpoint voltage. If it is outside these thresholds, then it is considered a valid logic level. Comparator results can also be ignored using a mask. Thus, there are typically three drive states (HI, LO, and HIZ) and five compare states (HI, LO, and MID, VALID, and MASK).

The usefulness of the valid comparison is not immediately obvious. If we want to test for valid VOH and VOL voltages from the output of a nondeterministic circuit such as an ADC, we cannot set the tester to expect HI or LO. This is because electrical noise in the ADC and tester will produce somewhat unpredictable results at the ADC output. However, we can set the tester to expect valid logic levels during the appropriate digital vectors without specifying whether the ADC should produce a HI or a LO. While the pin card tests for valid logic levels, the samples from the ADC are collected into the digital capture memory for later analysis.

In addition to the drive and compare circuits, digital pin cards may also include dynamic load circuits. A dynamic load is a pair of current sources connected to the DUT output with a diode bridge circuit as shown in the above figure. The diode bridge forces a programmable current into the DUT output whenever its voltage is below a programmable threshold voltage, V_{TH}. It forces current out of the DUT output whenever its voltage is above V_{TH}.

The sink and source current settings correspond to the DUT's IOH and IOL specifications. Another extremely important function that a digital pin card provides is its per-pin measurement capability. The per-pin measurement circuits of a pin card form a low-resolution, low-current DC voltage/current source for each digital pin.

The per-pin circuits also include a relatively low-resolution voltage/current meter. The low-resolution and low-current capabilities are usually adequate for performing certain DC tests like continuity and leakage testing. These DC source and measure circuits can also be used for other types of simple DC tasks like input or output impedance testing.

Some testers may also include overshoot suppression circuits that serve to dampen the overshoot and undershoot characteristics in rapidly rising or falling digital signals. The overshoot and undershoot characteristics are the result of a low-impedance DUT output driving into the DIB traces and coaxial cables leading to the digital pin card electronics. The ringing is minimized as the signal overshoot is shunted to a DC level through a diode.

Digital pin cards also include relays connected to other tester resources such as calibration standards and system DC meters and sources. These connections can be used for a variety of purposes, including calibration of the pin card electronics during the tester's system calibration process. The exact details of these connections vary widely from one tester type to another.

Timing and Formatting Electronics

When looking at a digital pattern for the first time, it is easy to interpret the ones and zeros very literally, as if they represent all the information needed to create the digital waveforms. However, most ATE testers apply timing and formatting to the ones and zeros to create more complicated digital waveforms while minimizing the number of ones and zeros that must be stored in pattern memory.

Timing and formatting is a type of data compression and decompression. The pattern data are formatted using the ATE tester's formatter hardware, which is typically located inside the tester mainframe or on the pin card electronics in the test head. Figure below shows how the pattern data are combined with timing and formatting information to create more complex waveforms.

Notice that the unformatted data in the below figure require four times as much 1/0 information and four times the bit cell frequency to achieve the same digital waveform as the formatted data. Another key advantage to formatted waveforms is that the formatting hardware in a high-end mixed-signal tester is capable of placing the rising and falling edges with an accuracy of a few tens of picoseconds. This gives us better control of edge timing than we could expect to achieve using subgigahertz clocked digital logic.

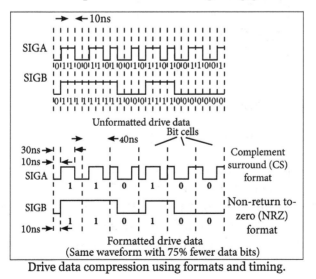

Drive data compression using formats and timing.

The programmable drive start and stop times illustrated in the above figure are generated using digital delay circuitry inside the formatter circuits of the tester. Drive and compare timing is refined during a calibration process called deskewing. This allows subnanosecond accuracy in the placement of driven edges and in the placement of compare times.

Strobe comparisons are performed at a particular point in time, while window comparisons are performed throughout a period of time. Window timing is typically used when comparing DUT outputs against expected patterns, while strobe timing is typically used when collecting data into capture memory. Again, this depends on the specific tester.

Figure below shows examples of several different formatting and timing combinations that create many different waveforms from the same digital data stream. In each case, the drive data sequence is 110X00. The compare data sequence is HHLXLL. Notice that certain formats such as Clock High and Clock Low ignore the pattern data altogether.

Since digital pin cards can both drive and expect data, a distinction is made between a driven signal (1 or 0) and an expected signal (H or L). In fact, some digital pattern standards define H/L as driven data and 1/0 as expected data.

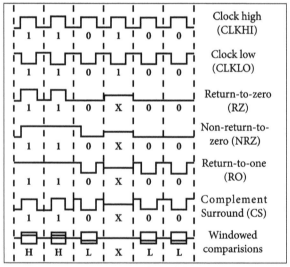

Some common digital formats.

3.6.3 AC Source and Measurement

AC Continuous-Wave Source and AC Meter

The simplest way to apply and measure single-tone AC waveforms is to use a continuous-wave source (CWS) and an RMS voltmeter. The CWS is simply set to the desired frequency and voltage amplitude to stimulate the DUT. The RMS voltmeter is equally simple to use. It is connected to the DUT output, and the RMS output is measured with a single test program command. But the CWS and RMS voltmeter suffer from a few problems.

First, they are only able to measure a single frequency during each measurement. This would be acceptable for bench characterization, but in production testing it would lead to unacceptably long test times. DSP-based multitone testing is a far more efficient way to test AC performance because multiple frequencies can be tested simultaneously.

Another problem that the RMS voltmeter introduces is that it cannot distinguish the DUT's signal from distortion and noise. Using DSP-based testing, these various signal components can easily be separated from one another. This ability makes DSP-based testing more accurate and reliable than simple RMS-based testing. DSP-based testing is made possible with a more advanced stimulus/measurement pair, the arbitrary waveform generator and the waveform digitizer.

Arbitrary Waveform Generators

An arbitrary waveform generator (AWG) consists of a bank of waveform memory, a DAC that converts the waveform data into stepped analog voltages, and a programmable low-pass filter section, which smoothes the stepped signal into a continuous waveform. An AWG usually includes an output scaling circuit (PGA) to adjust the signal level. It may also include differential outputs and DC offset circuits.

Figure below shows a typical AWG and waveforms that might be seen at each stage in its signal path. An AWG is capable of creating signals with frequency components below the low-pass filter's cutoff frequency. The frequency components must also be less than one-half the AWG's sampling rate.

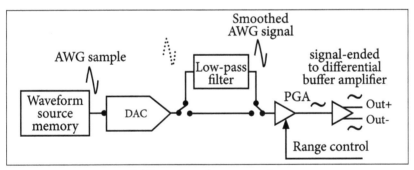

Arbitrary waveform generator.

An AWG might create the three-tone multitone as illustrated in the below figure. It might also be used to source a sine wave for distortion testing or a triangle wave for ADC linearity testing. Flexibility in signal creation is the main advantage of AWGs compared to simple sine wave or function generators.

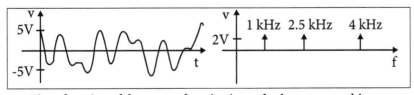

Time-domain and frequency-domain views of a three-tone multitone.

Waveform Digitizers

An AWG converts digital samples from a waveform memory into continuous-time waveforms. A digitizer performs the opposite operation, converting continuous-time analog waveforms into digitized representations. The digitized samples of the continuous waveform are collected into a waveform capture memory.

The structure of a typical digitizer is shown in the below figure. A digitizer usually includes a programmable low-pass filter to limit the bandwidth of the incoming signal. The purpose of the bandwidth limitation is to reduce noise and prevent signal aliasing.

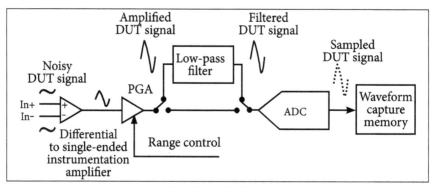

Waveform digitizer.

Like the DC meter, the digitizer has a programmable gain stage at its input to adjust the signal level entering the digitizer's ADC stage. This minimizes the noise effects of quantization error from the digitizer's ADC. Waveform digitizers may also include a differential to single-ended conversion stage for measuring differential outputs from the DUT.

Digitizers may also include a sample-and-hold circuit at the front end of the ADC to allow under-sampled measurements of very high-frequency signals.

Clocking and Synchronization

Many of the subsections and instruments in a mixed-signal tester derive their timing from a central frequency reference. This frequency determines the repetition rate of the sample loop and therefore sets the frequency of the DAC or ADC sampling rates. The AWG and digitizer also operate from clock sources that must be synchronized to each other and to the digital pattern's frame loop repetition rate.

Figure below shows a clock distribution scheme that allows synchronized sampling rates between all the DSP-based measurement instruments. Since the clocking frequency for each instrument is derived from a common source, frequency synchronization is possible. Without precise sampling rate synchronization, the accuracy and repeatability of all the DSP-based measurements in a mixed-signal test program would be degraded.

Proper synchronization of sample rates between the various AWGs, digitizers and digital pattern generators is another of the key distinguishing features of a mixed-signal tester. A digital tester with bolt-on analog instruments often lacks a good clocking and synchronization architecture.

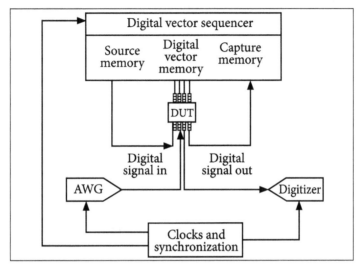

Synchronization in a mixed-signal tester.

3.6.4 Time Measurement System

Time Measurements

Digital and mixed-signal devices often require a variety of time measurements, such as frequency, period, duty cycle, rise and fall times, jitter, skew and propagation delay. These parameters can be measured using the ATE tester's time measurement system (TMS). Most TMS instruments are capable of measuring these parameters within an accuracy of a few nanoseconds. Some of the more advanced TMS instruments can measure parameters such as jitter to a resolution of less than 1 ps.

Timing parameters that do not change from cycle to cycle (i.e., rise time, fall time, etc.) can sometimes be measured using a very high-bandwidth under sampling waveform digitizer. An under sampling digitizer is similar in nature to the averaging mode of a digitizing oscilloscope.

Like digitizing oscilloscopes, under sampling digitizers require a stable, repeating waveform. Thus, non-periodic features such as jitter and random glitches cannot be measured using an under sampling approach. Unfortunately, under sampling digitizers are often considerably slower than the dedicated time measurement instruments.

Time Measurement Interconnects

One of the most important questions to consider about a TMS instrument is how its input and interconnection paths affect the shape of the waveform to be measured.

It does little good to measure a rise time of 1 ns if the shape of the signal's rising edge has been distorted by a 50-Ω coaxial connection. It is equally futile to try to measure a 100-ps rising edge if the bandwidth of the TMS input is only 300 MHz. Accurate timing measurements require a high-quality signal path between the DUT output and the TMS time measurement circuits.

3.6.5 Computing hardware

User Computer

Mixed-signal testers typically contain several computers and signal processors. The test engineer is most familiar with the user computer, since this is the one that is attached to the keyboard. The user computer is responsible for all the editing and compiling processes necessary to debug a test program. It is also responsible for keeping track of the datalogs and other data collection information.

On low-cost testers, the user computer may also drive the measurement electronics as well. On more advanced mainframe testers, the execution of the test program, including I/O functions to the tester's measurement electronics, may be delegated to one or more tester computers located inside the tester's mainframe.

Tester Computer

The tester computer executes the compiled test program and interfaces to all the tester's instruments through a high-speed data backplane. By concentrating most of its processing power on the test program itself, the tester computer can execute a test program more efficiently than the user computer.

The tester computer also performs all the mathematical operations on the data collected during each test. In some cases, the more advanced digital signal processing (DSP) operations may be handled by a dedicated array processor to further reduce test time. However, computer workstations have become fast enough in recent years that the DSP operations are often handled by the tester computer itself rather than by a dedicated array processor.

Array Processors and Distributed Digital Signal Processors

Many mixed-signal testers include one or more dedicated array processors for performing DSP operations quickly. This is another difference between a mixed-signal tester and a bolted-together digital/analog tester. Some mixed-signal instruments may even include local DSP processors for computing test results before they are transferred to the tester computer. This type of tester architecture and test methodology is called distributed processing.

Distributed processing can reduce test time by splitting the DSP computation task among several processors throughout the tester. Test time is further reduced by eliminated much of the raw data transfer that would otherwise occur between digitizer instruments and a centralized tester computer or array processor. Unfortunately, distributed processing may have the disadvantage that the resulting test code may be harder to understand and debug.

Network Connectivity

The user computer and/or tester computer are typically connected into a network using Ethernet or similar networking hardware. This allows data and programs to be quickly transferred to the test engineer's desk for offline debugging and data analysis. It also allows for large amounts of production data to be stored and analyzed for characterization purposes.

3.7   IDDQ Testing

Applying a power supply voltage to a CMOS chip causes a current IDD to flow. When the signal inputs are stable (not switching), the quiescent leakage current IDDQ can be measured. This is illustrated in the below figure. Every chip design is found to have a range of 'normal' levels. IDDQ testing is based on the assumption that an abnormal reading of the leakage current indicates a problem on the chip. IDDQ testing is usually performed at the beginning of the testing cycle. If a die fails, it is reflected and no further tests are performed.

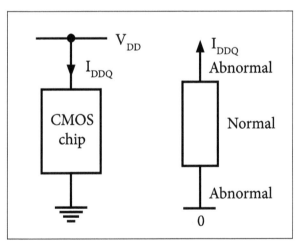

Basic IDDQ test.

The components of a basic IDDQ measurement system is shown in below Figure. The test chip is modeled as being in parallel with the testing capacitance, C_{test} . A power supply with a value V_{DD} is connected to the chip by a switch that is momentarily closed at time t = 0. The current IDD is monitored by a buffer (a unity-gain amplifier) and gives the output voltage.

$$I_{DD} = C[\Delta V_o / \Delta t]$$

Where the voltage falls by an amount ΔV_o in a time Δt.

The total capacitance C in the equation is the sum and can be expressed as:

$$C = C_{test} + C_{chip}$$

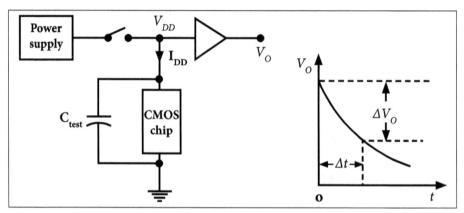

Components of an IDDQ measurement system.

3.7.1 Design for Testability

It is an absolute requirement for Modern Chips. Partition the chip to make controllability of all state and therefore testing becomes easier.

- Design for testability shows the fault model to test the chip in manufacturing process.

- The aspect of testability is based on two key concepts i.e., controllability and observability. The inclusion of these concepts means the provision of some means for setting and resetting key nodes in the system and then observing the response at key points in the system.

Design for testability is then just a set of design rules or guidelines, if obeyed will facilitate the test. Any failure occurs during testing is due to either poorly controlled fabrication process or because of design defect.

Wide Range of Options

- Explicit built-in-self-test (BIST).

- Additions to the (micro) code to make it go through a test sequence.

- Extra test buses.

- Link all the registers/latches into scan chains, to make state scannable.

There are some automatic test pattern generation programs that do a really good job, but only if all state is scannable. This is a motivation to make the state scannable even if it seems excessive.

Technology Options, Power Calculations

Instantaneous Power

The instantaneous power P(t) drawn from the power supply is proportional to the supply current $i_{DD}(t)$ and the supply voltage V_{DD},

$$P(t) = i_{DD}(t)V_{DD}$$

Energy

The energy consumed over the time interval T is the integral of P(t):

$$E = \int_0^T i_{DD}(t)V_{DD}dt$$

Average Power

The average power over this interval is given by:

$$P_{avg} = \frac{E}{T} = \frac{1}{T}\int_0^T i_{DD}(t)V_{DD}dt$$

Analog Circuit Design for Testability

Analog circuit testing differs from both logic and memory testing. In short, analog circuits are tested for their specifications, which are expressed in terms of functional parameters, such as voltage levels, frequency response, inter-modulation distortion, etc.

Here is an incomplete list of analog blocks, often found in mixed-signal devices RIM analog switches, analog-to-digital (AID) converters, comparators, digital-to-analog (D/A) converters, operational amplifiers, precision voltage references, phase-locked loops, resistors, capacitors, inductors and transformers.

There are two types of measurements associated with analog circuit testing. In the first type, we measure component values and ascertain that they are within the specified range. In the other type of measurement, signal characteristics are measured. These require applying carefully generated signal waveforms to the inputs and analyzing the outputs. Both the functions are generally accomplished by digital signal processing (DSP) techniques.

The main idea of design for testability (DFT) for analog circuits is to provide access to selected nodes for testing. Traditionally, signal generation and DSP were provided by the expensive automatic test equipment (ATE.) With the current trend of integrating analog and digital circuitry on the same chip, often the DSP function is available on-chip. On-chip digital functions have also been used for creating self-test for the analog portion.

3.7.2 Built-In Self-Test

Built-In Self-Test (BIST) is a concept that a chip can be provided with the capability to test itself. There are several ways to accomplish this objective. One way is that the chip tests itself during normal operation. In other words, there is no need to place the chip under test into a special test mode. We call this the on-line BIST.

We can further divide on-line BIST into concurrent on-line BIST and non-concurrent on-line BIST. Concurrent on-line BIST performs the test simultaneously with normal functional operation. This is usually accomplished with coding techniques (e.g., parity check). Non-concurrent BIST performs the test when die chip is idle.

Off-line BIST tests the chip when it is placed in a test mode. An on-chip pattern generator and a response analyzer can be incorporated into the chip to eliminate the need for external test equipment. Test patterns developed for a chip can be stored on the chip for BIST purposes. However, the storage of a large set of test patterns increases the chip area significantly and is impractical.

A pseudo-random test is carried out instead. In a pseudo-random test, pseudo-random numbers are applied to the circuit under test as test patterns and the responses compared to expected values.

A pseudo-random sequence is a sequence of numbers that is characteristically very similar to the random numbers. However, pseudo-random numbers are generated mathematically and are deterministic. In this way, the expected responses of the chip to these patterns can be predetermined and stored on chip.

The storage of the chip's correct responses to pseudo-random numbers also has to be avoided for the same reason of avoiding the storage of test patterns. An approach called signature analysis was developed for this purpose. A component called a signature register can be used to compress all the responses into a single vector (signature) so that the comparison can be done easily. Signature registers are also based on linear feedback shift registers.

Linear Feedback Shift Register

Linear Feedback Shift Register (LFSR) are used in GIST both as a generator of pseudo-random patterns and as a compressor of responses. The below Figure shows signature analyzer consisting of the feedback shift register, which illustrates the sequence it generates. Each box represents a flip-flop. The flip-flops are synchronized by a common clock and form a rotating shift register. Assume that the initial value in the shift register is 110. It is shown that the shift register goes through a 3-pattern sequence -110-011-101. The sequence repeats afterward.

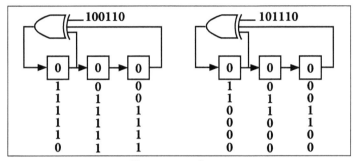

Signature analyzer.

When a sequence of n bits is encoded by m-bit signature (m < n), more than one sequence will map into one signature. There are 2n unique sequences and 2m unique signatures in this situation. In average, each signature will represent $2^n / 2^m = 2^{n-m}$ sequences. The probability of declaring an incorrect sequence correct since it produces the expected signature is,

$$\frac{2^{n-m} - 1}{2^n - 1} \qquad \qquad \dots(1)$$

The denominator in (2) is the number of incorrect sequences. The numerator is the number of incorrect sequences that would map into the signature identical with that of the correct sequence. Normally n >> m > 1, so (1) can be approximated as,

$$\frac{2^{n-m} - 1}{2^n} = 2^{-m} \qquad \qquad \dots(2)$$

The probability of drawing an incorrect conclusion from using a signature analyzer can then be made arbitrarily small by choosing a large m. Normally, m = 16 would give an acceptable error probability. When the signature is incorrect, the circuit is not functioning properly. If the signature is correct, the circuit has a high probability to be functioning correctly.

Multiple data sequences can be combined and compressed with a signature analyzer with multiple inputs to produce a multiple-input signature. Below Figure shows the use of a pseudo-random pattern generator and a signature analyzer to test a circuit.

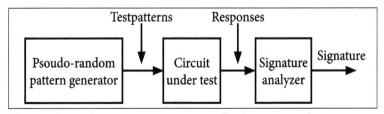

Set-up of a pseudo-random pattern generator and a signature analyser to test a circuit.

Finite State Machine Approach for BIST

Flip-flops have two main uses within circuits. Firstly, they are used to store logic values (or more commonly, a group of flip-flops is used as a data register to store a logic word) for use at some later stage in the process. In this kind of application, testing will often be relatively straightforward, since the inputs and outputs are likely to be reasonably accessible and the relationship between input and output is uncomplicated.

The other main use for flip-flops is as the central components in Finite State Machines (FSM's). A FSM is used to control the execution of a sequence of operations, this is achieved by making each operation depend on a state of the FSM, where a state is defined as a particular set of values held in its flip-flops.

The FSM changes state under the control of a clock, but the particular sequence of states that it passes through is defined by the signals applied to the inputs of the flip-flops. These signals will be generated by a clock of combinational circuitry the next-state logic. If the next state depends only on the present state, the FSM has no external inputs (apart from the clock). It produces a fixed sequence of states and is known as an autonomous FSM.

In general, however, a FSM can have external inputs which modify its behavior so that the state transition at any time is a function both of the present state and of the external inputs. Such a machine can be represented as in the below Figure, which shows an FSM with two flip-flops, X and Y and two external inputs, A and B. If the flip-flops are, for the sake of argument, D type, then the next state logic has to produce flip-flop input signals D_x and D_y as functions of A, B, X and y.

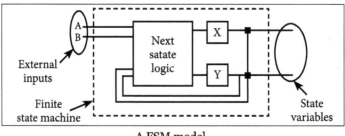

A FSM model.

The requirements for this logic can be expressed in terms of a state transition diagram, an example of which is shown in the below Figure. There are several features of this diagram that have a bearing on testing activities and problems.

- A FSM with n flip-flops and m external inputs will contain 2n states and 2m transitions per state. Representing all these states and transitions quickly becomes unwieldy to the point of incomprehensibility as the circuit increases in size; this is equally true whether diagrammatic or tabular methods are used.

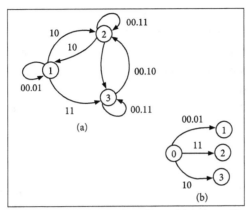

State-transition diagram: (a) System requirement for all combination of inputs and outputs (b) Fourth state with transitions.

- When designing a FSM, the state-transition diagram will be derived from the specification. In particular, the number of states required depends on the application and can take any integral value. When it comes to implementation, the number of flip-flops in the FSM must be chosen so that there are enough states available; this will often mean that the design contains redundant states. In the above figure (a), for example, three states are specified, the FSM must, therefore, contain two flip-flops, which means that four states will actually exist. An implementation of the FSM of Figure (a) is shown in below Figure. In deriving this, transitions from state 0 are entered as 'don't care'. However, once the circuit has been implemented then the logic that is designed to produce the required transitions among the 'working' states will also of necessity define transitions from the redundant state these are shown in above Figure (b).

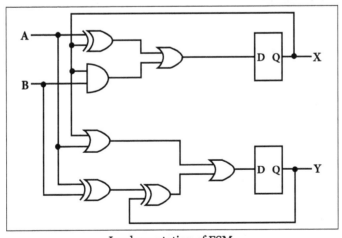

Implementation of FSM.

The state-transition diagram gives no indication of how the circuit will be have when first switched on. In fact, unless special measures are taken, the circuit can settle entirely unpredictably into any one of the possible states, including any of the redundant states. This indeterminacy has to be allowed for both in the functional design and when testing. Working from the circuit diagram, even from the very simple example of above Figure, the function of the circuit is far from clear. In particular, there is no way in which the working states can be distinguished from redundant ones. Testing has to be developed largely on a structural basis using the circuit diagram. The only real alternative would be a hybrid approach, treating individual flip-flops on a functional basis while using structural methods for the 'glue' logic.

Embedded State Machines

The problems posed by FSM's circuits are further increased if it is embedded within further blocks of logic so that its behavior can only be inferred by observation of output values. An example of an embedded FSM is shown in below Figure, which represents an autonomous FSM whose state variables provide the inputs to a block of output logic which forms the single output variable W.

The waveform can be seen to have a period of five clock cycles and can, therefore, be generated by a five-state FSM. The output is required to be high during states 3 and 4, with states 0, 6 and 7 being redundant. Using these redundancies, we can form the function as:

$$W = YZ + \overline{Y}.\overline{Z}$$

The existence of combinational logic between the state variables and the primary outputs can have a number of consequences:

- The fault cover for any particular test will probably be reduced.

- To establish a sensitive path through the output logic will require a particular state, which may require a sequence of input patterns.

- Some faults may well become untestable.

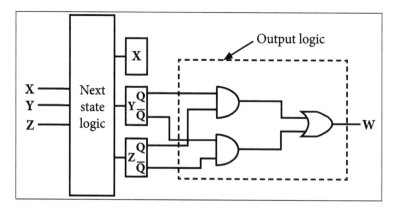

A typical waveform generator having a FSM with logic output.

3.7.3 Boundary Scan

Mixed-Signal Boundary Scan (IEEE Std. 1149.4)

The IEEE Std. 1149.4 mixed-signal boundary scan standard was developed by many companies and academic institutions around the world. IEEE 1149.4 is built upon the 1149.1 digital boundary scan standard. As an analog complement to the 1149.1 boundary scan for digital circuits, the 1149.4 standard allows chip-to-chip interconnect testing of analog signals.

Optionally, it allows testing of internal circuit nodes. The 1149.4 standard provides a consistent interface for analog and mixed-signal tests for those signals that can tolerate the loading, series resistance, and cross-talk issues inherent in the physical implementation of the standard in the target IC process (e.g., CMOS).

The 1149.4 mixed-signal boundary scan standard is compliant with the 1149.1 digital TAP and boundary scan architecture. The major difference between 1149.4 and 1149.1 is that the 1149.4 standard includes some new test pins and analog switches for exercising non digital circuits. Figure below shows the analog boundary module (ABM) for the 1149.4 standard.

The ABM provides standardized access to analog input and output signals at the external device pins. The 1149.4 standard allows a simple chip-to-chip interconnect verification scheme similar to that used in traditional digital boundary scan. A pair of switches at each analog input and output pin of the IC allows the pin's normal (analog) signal to be replaced by digital signal levels, VH and VL.

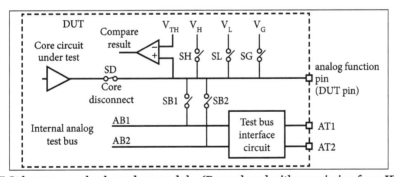

IEEE Std. 1149.4 analog boundary module. (Reproduced with permission from IEEE.)

V_H and V_L would typically be connected to VDD and digital ground. In effect, the analog input or output becomes a simple digital driver. The interconnect between ICs can be tested by forcing either V_H or V_L from the pin and then checking the status of a receiver at the other end of the interconnection. The receiver, also part of the 1149.4 standard is an analog comparator tied to an 1149.1 digital boundary scan cell. It compares the incoming voltage against a threshold voltage, VTH. In addition to the two logic level connections, the analog pin can also be connected to a quality ground, V_G (typically analog ground).

Although the 1149.4 standard is primarily targeted for chip-to-chip interconnects testing, it does include optional extensions for internal analog signal testing. For this purpose, the 1149.4 standard uses a pair of analog test buses, similar in nature to the analog test input and output buses. The analog switches are controlled by shifting control bits into the 1149.1 TAP, allowing a standardized method of setting up analog stimulus and measurement interconnects.

The analog buses can be used for a variety of purposes; including internal testing as well as external (chip-to-chip) interconnects testing. As an example of external testing, engineers at Hewlett Packard and Ford Motor Company developed a method to use this structure to verify the interconnects between ICs and networks of passive components such as resistors, capacitors, and inductors. This method forces DC or AC current through one test bus while measuring the voltage response through the other bus.

It should be noted that the switches defined by the 1149.4 standard do not necessarily need to be physical switches. For example, if the output of a particular circuit can be set to a high-impedance state, then it does not need to be disconnected using a switch. Similarly, if a circuit's output can be set to force a high level and a low level under 1149.1 digital control, then separate VDD and ground switches are not needed.

The switches defined by the standard are therefore behavioral in nature, rather than physical requirements. The advantage of eliminating switches when possible is two-fold. First, the series impedance and/or capacitive loading of a CMOS transmission gate or other switching structure is not introduced into the signal path. Second, the silicon required to implement the 1149.4 standard can be minimized if the number of switches can be minimized.

The 1149.4 standard cannot be employed blindly to test internal signals without consideration of the effects of the standard on the analog circuits to be measured. Actually, it is not the standard that is the problem. It is the practical implementation of the standard using CMOS or other types of analog switches. Signal crosstalk, capacitive loading, and increased noise and distortion are possible problems that may occur when using CMOS switches in sensitive analog circuits.

In some cases, the design engineer might need to use T-switch configurations to minimize signal crosstalk and injected noise, though the 1149.4 standard does not specify the physical embodiment of the switches. The issues of crosstalk, noise injection, and loading are identical to those in the more general ad hoc mixed-signal test bus configurations, which have been used successfully for many years.

The problems are not insurmountable. They simply require the design engineer to evaluate which nodes can and cannot tolerate the potential imperfections introduced by the analog switches. Like the 1149.1 standard, the 1149.4 standard carries more overhead than the traditional ad hoc methods. But like the 1149.1 standard, the extra baggage is well justified by the tremendous enhancement in the standardization of test access. For the same reasons outlined in the 1149.1 section, the overhead will eventually be much less of a problem as processing geometries shrink.

Scan Design

Scan design is implemented to provide controllability and observability of internal state variables for testing a circuit.

It is also effective for circuit partitioning.

A scan design with full controllability and observability turns the sequential test problem into a combinational one.

Scan Design Requirements

Flip-flops are replaced by scan flip-flops (SFF) and are connected so that they behave as a shift register in the test mode. The output of one SFF is connected to the input of next SFF. The input of the first flip-flop in the chain is directly connected to an input pin (denoted as SCANIN) and the output of the last flipflop is directly connected to an output pin (denoted as SCANOUT). In this way, all the flip-flops can be loaded with a known value and their value can be easily accessed by shifting out the chain. The Figure shows a typical circuit after the scan insertion operation.

- Input/output of each scan shift register must be available on PI/PO.

- Combinational ATPG is used to obtain tests for all testable faults in the combinational logic.

- Shift register tests are applied and ATPG tests are converted into scan sequences for use in manufacturing test.

Unfortunately, there are two types of overheads associated with this technique that the designers care about very much. They are the hardware overhead and performance overhead.

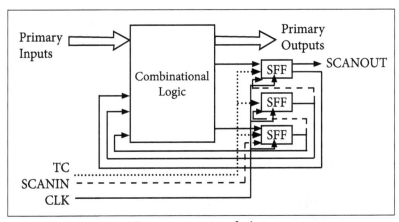

Scan structure to a design.

There have been many variations of scan. They are:

- MUX-ed Scan.

- Scan path.

- Scan-Hold Flip-Flop.

- Serial scan.

- Level-Sensitive Scan Design (LSSD).

- Scan set.

- Random access scan.

MUX-ed Scan

In this approach, a MUX is inserted in front of each FF to be placed in the scan chain. The below figure shows that when the test mode pin T = 0, the circuit is in normal operation mode and when T = 1, it is in test mode (or shift-register mode). The scan flip-flips (FF's) must be interconnected in a particular way. This approach effectively turns the sequential testing problem into a combinational one and can be fully tested by compact ATPG patterns.

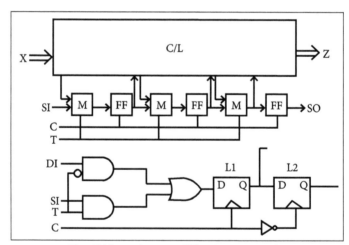

The Shift-Register Modification approach.

Scan Path

This approach is also called the Clock Scan Approach. It was invented by Kobayashi et al. in 1968 and reported by Funatsu et al. in 1975 and adopted by NEC. In this approach, multiplexing is done by two different clocks instead of a MUX. It uses two-port race less D-FF's as shown in Figure below. Each FF consists of two latches operating in a master-slave fashion and has two clocks (C_1 and C_2) to control the scan input (SI) and the normal data input (DI) separately.

The two-port race less D-FF is controlled in the following way:

- For normal mode operation, C_2 = 1 to block SI and C_1 = 0 → 1 to load DI.

- For shift register test mode, C_1 = 1 to block DI and C_2 = 0 → 1 to load SI.

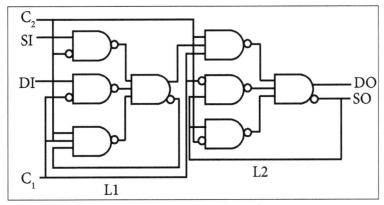

Logic diagram of the two-port race less D-FF.

This approach gives a lower hardware overhead (due to dense layout) and less performance penalty (due to the removal of the MUX in front of the FF) compared to the MUX Scan Approach. The real figures however depend on the circuit style and technology selected and on the physical implementation.

Level-Sensitive Scan Design (LSSD)

This approach was introduced by Eichelberger and T. Williams in 1977 and 1978. It is a latch-based design used at IBM. It guarantees race-free and hazard-free system operation as well as testing. It is insensitive to component timing variations such as rise time, fall time and delay. It is faster and has a lower hardware complexity than SR modification. It uses two latches (one for normal operation and one for scan) and three clocks.

Furthermore, the designer has to follow a set of complicated design rules. A logic circuit is level sensitive (LS) if the steady state response to any allowed input change is independent of the delays within the circuit. Also, the response is independent of the order in which the inputs change.

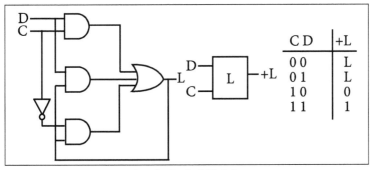

A polarity-hold latch.

LSSD requires that the circuit be LS, so we need LS memory elements as defined above. The above figure shows an LS polarity-hold latch. The correct change of the latch output (L) is not dependent on the rise/fall time of C, but only on C being '1' for a period of time greater than or equal to data propagation and stabilization time. The below figure shows the polarity-hold shift-register latch (SRL) used in LSSD as the scan cell.

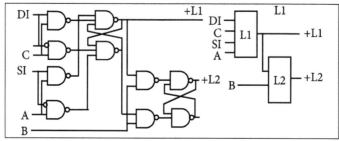

The polarity-hold shift-register latch (SRL).

The scan cell is controlled in the following way:

Normal mode: A = B = 0, C = 0 → 1

SR (test) mode: C = 0, AB = 10 → 01 to shift SI through L1 and L2

Random Access Scan

This approach was developed by Fujitsu and was used by Fujitsu, Amdahl and TI. It uses an address decoder. By using address decoder we can select a particular FF and either set it to any desired value or read out its value. The below figures shows a random access structure and the RAM cell:

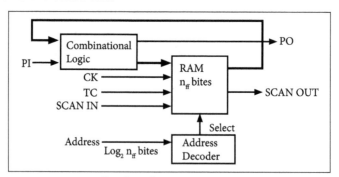

Random Access structure.

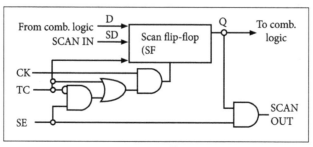

RAM cell.

The difference between this approach and the previous ones is that the state vector can now be accessed in a random sequence. Since neighboring patterns can be arranged so that they differ in only a few bits and only a few response bits need to be observed, the test application time can be reduced.

In this approach, test length is reduced. This approach provides the ability to 'watch' a node in normal operation mode, which is impossible with previous scan methods. This is suitable for delay and embedded memory testing. The major disadvantage of the approach is high hardware overhead due to address decoder, gates added to SFF, address register, extra pins and routing.

Scan-Hold Flip-Flop

Special type of scan flip-flop with an additional latch designed for low power testing application. The below figure shows a hold latch cascaded with the SFF. The control input HOLD keeps the output steady at previous state of flip-flop.

For HOLD = 0, the latch holds its state and for HOLD = 1, the hold latch becomes transparent. For normal mode operation, TC = HOLD = 1 and for scan mode, TC = 1 and Hold = 0.

Hardware overhead increases by about 30% due to extra hardware the hold latch. This approach reduces power dissipation and isolate asynchronous part during scan. It is suitable for delay test.

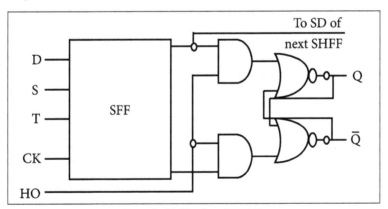

Scan-hold flip-flop (SHFF).

Partial Scan Design

In this approach only a subset of flip-flops is scanned. The main objectives of this approach are to minimize the area overhead and scan sequence length. It would be possible to achieve required fault coverage.

In this approach sequential ATPG is used to generate test patterns. Sequential ATPG has number of difficulties such as poor initialization, poor controllability and observability of the state variables etc. Number of gates, number of FF's and sequential depth give little idea regarding testability and presence of cycles makes testing difficult. Therefore, sequential circuit must be simplified in such a way so that the test generation becomes easier.

Removal of selected flip-flops from scan improves the performance and allows limited scan design rule violations. It also allows automation in scan flip-flop selection and test generation The below figure shows a design using partial scan architecture. Sequential depth is calculated as the maximum number of FF's encountered from PI line to PO line.

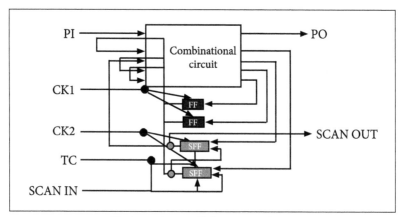

Design using partial scan structure.

3.7.4 Analog Test Bus, System Test and Core Test

The analog test bus (ATB) (IEEE Standard 1149.4) architecture is illustrated in the below figure for a mixed-signal device [214, 387, 405, 435, 510, 511, 554, 658, 719].

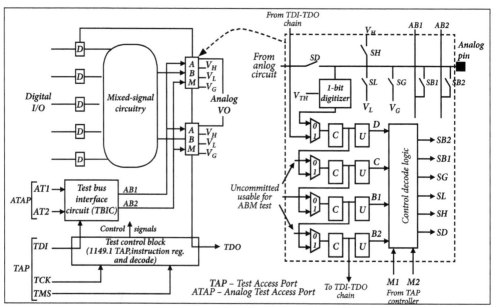

Analog test bus architecture.

This bus replaces the in-circuit tester, because it provides electronic access and multiplexing hardware in order to probe various digital and analog components in a mixed-signal chip or various external impedances connected to the pins of the chip. We use the 1149.4 analog test bus [319] with the JTAG 1149.1 digital boundary scan standard [318] to add testability to test some analog circuits.

Its disadvantages are as follows:

- We eliminate a large analog chip area needed for extra test points.

- We gain analog circuit observability, particularly at the interface between the analog and digital parts of the chip.

- The C-switch sampling devices couple all probe points capacitively, even when the test bus is not in use. To avoid this unwanted change in the analog circuit transfer function, we use more elaborate analog C-switches in the 1149.4 standard.

- The bus may have a 5% measurement error.

- There is a stringent limit on how far one can move analog data through a system using the 1149.4 standard before having to digitize it to preserve accuracy.

Emerging Technology and IC Design for Internet of Everything (IoE)

4.1 Overview of LDMOS

LDMOS (Laterally Diffused Metal Oxide Semiconductor) transistors are voltage controlled devices, hence unlike the bipolar devices, there is no gate current. Hence, the bias circuitry is very much simplified as compared to the bipolar devices. The majority of the LDMOS devices have the source connected to the backside of the device. Hence, the requirement of toxic Beryllium oxide (BeO) packages is eliminated. The bulk source can be eutectically soldered to the package and the bond wire requirement is removed, reducing the inductance.

The LDMOS devices show better temperature stability than bipolar devices. Also they provide device stabilization, which prevents oscillations at higher frequencies. A cross-section schematic of the LDMOS is shown in the below figure. The device has a sinker diffusion connecting the source to the backside substrate. The device consists of a drain extension, which helps in realizing higher breakdown voltages.

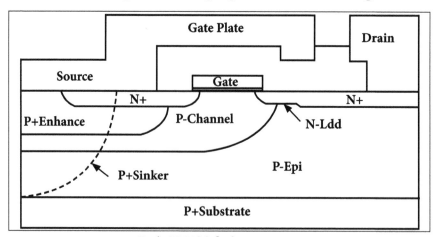

Basic LDMOS device structure.

The drain is shielded from the gate by metal field plate, realizing extremely low feedback capacitance. The higher breakdown value in LDMOS is due to the Reduced Surface Field technology (RESURF). In RESURF, one horizontal PN junction and a vertical p+n junction develop the two diode structures. The vertical diode will have a lower breakdown voltage, which can be determined by the epitaxial doping level. The horizontal junction breakdown voltage is higher due to the high ohmic substrate.

At thinner layers of the epitaxial layer, the depletion of the vertical junction becomes more and more reinforced by the horizontal junction. Hence for the same applied voltage, the depletion layer stretches longer along the surface than expected from one dimensional calculation. After a certain thickness, the reduced surface field does not reach the critical value even at high voltages and hence, the breakdown is eliminated or raised to very high voltages. This forms the basis of an increased breakdown voltage LDMOS device.

LDMOS Fabrication Process

The LDMOS fabrication process is given in the below table. The process starts with a conventional bulk silicon wafer. In Chipfilm™ technology, 1-2 µm thick wafer surface is the substrate. An epitaxial layer is grown over this layer. The buried p$^+$ doping profile of the Chipfilm™ wafers is replicated by simulating the epitaxial layer growth over a p$^+$ layer.

A p-epitaxial layer of thickness 2 µm is grown over the starting wafer. The epitaxial layer is p-doped with a boron concentration of 1×10^{15} cm^{-3}. A thin layer of gate oxide of thickness 57.3 nm is grown using dry oxidation process. A boron threshold voltage adjust implantation is carried out next.

LDMOS Device Fabrication Process Steps

Step	Description
1	Starting material initial p-type substrate
2	p-epitaxy deposition
3	Gate oxide deposition by dry oxidation
4	Gate oxide patterning and polysilicon gate deposition
5	Poly silicon gate patterning
6	Patterning for drain extension diffusion
7	N$^+$ diffusion for poly gate and drain extension region
8	Source and drain region patterning
9	N$^+$ diffusion for source and drain regions
10	Fermi compress anneal
11	Contact electrode area patterning and metal deposition

Next, the polysilicon gate is deposited and patterned. Also, the region for drain extension diffusion is patterned and an n$^+$ implant is carried out to dope the polysilicon gate and to produce a n-doped drain extension region besides the gate. The drain and source implant regions are patterned and n$^+$ implant is carried out. Finally, a fermi compress anneal followed by metal drain source contact deposition is carried out. The process parameters are given in the below table.

Fabrication Process Parameters

Step	Description
Epitaxy deposition	Boron conc = 1e14, 45 min, 900°C temperature
Dry oxidation	Duration 30 min, temperature 1000°C
V_t adjust implant	Boron implantation, dose = 6×10^{11}, Energy = 20 keV
N$^+$ diffusion for poly gate and drain extension region	Phosphorous dose = 2×10^{14}, Energy = 100 keV
Source drain diffusion	Phosphorous dose = 3×10^{15}, Energy = 100 keV

4.1.1 Power MOS

The conventional planar MOSFET has the restriction of handling high power. In high power applications, the double diffused vertical MOSFET or VMOS is used which is simply known as power MOSFET.

MOSFETs are the majority carrier device, which means the flow of current inside the device is carried out either by the flow of electrons (N-Channel MOSFET) or flow of holes (P-Channel MOSFET). So when the device turns off, reverse recombination process would not happen. It leads to short turn ON/OFF times. As switching time is less, loss associated with it is less.

Power MOSFET.

So for high frequency applications, where the switching loss is a major impact on total power loss of the circuit, this device is the right choice. On the other hand, for applications having lower operating frequencies, BJT is superior.

Power MOSFET is available in two basic forms:

1. Depletion type

The transistor requires the gate source voltage (V_{GS}) to switch OFF the device. The depletion mode MOSFET is equivalent to a Normally Closed switch.

2. Enhancement type

The transistor requires the gate source voltage (V_{GS}) to switch ON the device. The enhancement mode MOSFET is equivalent to a Normally Open switch.

Two basic types of MOSFET are:

- N-channel MOSFET

- P-channel MOSFET

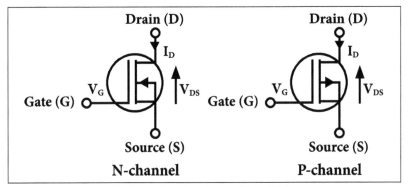

N-channel and P-channel MOSFET.

In N-channel MOSFET, the current conduction is due to the flow of electrons in the inversion layer, whereas in P-channel MOSFET, the current conduction is due to the flow of holes.

Working Principle of MOSFET

There are basically three regions in which MOSFETs can operate:

1. Cut-off region

In this region, MOSFET is in non-conducting state i.e., turned OFF and channel current IDS = 0. The gate voltage VGS is less than the amount of threshold voltage required for conduction.

2. Linear region

In this region, the channel is conducting and controlled by the gate voltage. For the MOSFET to be in this state, V_{GS} must be greater than the threshold voltage and also the voltage across the channel VDS must be greater than V_{GS}.

3. Saturation region

In this region, MOSFET is turned ON. The voltage drop for a MOSFET is typically lower than that of a bipolar transistor and as a result, powers MOSFETs are widely used for switching large currents.

Characteristics

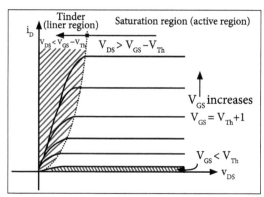

V-I characteristics of n-channel enhancement mode MOSFET.

Switching for Different Types of MOSFET

MOSFET type	$V_{GS}(+ve)$	$V_{GS}(0)$	$V_{GS}(-ve)$
N-channel enhancement	ON	OFF	OFF
N-channel depletion	ON	ON	OFF
P-channel enhancement	OFF	OFF	ON
P-channel depletion	OFF	ON	ON

4.1.2 Floating Gate MOS

The figure below shows a Floating Gate MOSFET (FGMOS), which is essentially present in a flash memory cell. There are two gates in FGMOS. The second gate is known as the floating gate, because it is completely electrically isolated.

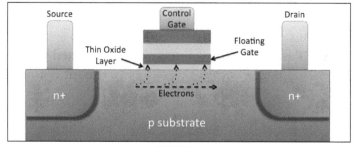

FGMOS.

In the above figure note that, the oxide layer beneath the floating gate is deliberately thinner than that above it.

In a FGMOS, if a high charge is applied to the control gate in the same manner as with a MOSFET, electrons flowing from source to drain can get excited and jump through the oxide layer into the floating gate, thus increasing its retained charge. The floating gate is the bucket of electrons.

To erase the charge stored on the floating gate, a high voltage is applied across the source and drain while a negative voltage is applied to the control gate, causing the retained electrons to jump back off the floating gate. The word jump is slightly complicated and usually involves a process called Fowler-Nordheim tunnelling. It is complicated, because it involves quantum mechanics.

Read operations

Now that we have methods for programming and erasing, we just need a way of testing the value stored i.e., a read operation. In MOSFET, we could control the flow of charge between the source and drain by varying the voltage applied to the gate.

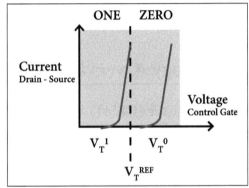

FGMOS read thresholds.

The voltage threshold at which current begins to flow from drain to source is different depending on the charge stored on the floating gate. By testing at an intermediate reference voltage $\left(V_T^{REF}\right)$ called the read point we can determine whether the floating gate contains charge (which we call ZERO) or not (which we call ONE).

In FGMOS, this can be turned around so that, by measuring the current we can determine the voltage on the floating gate, because electrons trapped on the floating gate cause the threshold. By applying a certain voltage $\left(V_T^{REF}\right)$ across the source and drain and then testing the current, we can determine if the voltage on the gate is above or below a specific point, called the read point.

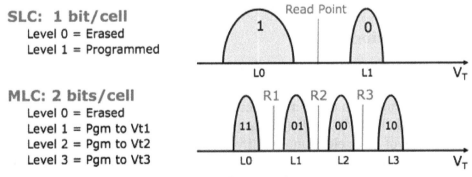

Read points for SLC and MLC.

The cell configuration of FGMOS comes in different forms, such as SLC, MLC and TLC. MLC reads slower than SLC and TLC even slower. This is because SLC contains only one bit of data, so we need to test one threshold voltage i.e., SLC has only one read point, but for MLC, where there are two bits, there are three read points, while for TLC there are even more.

When we test the SLC bucket to see if it is above or below 50% full, the result will tell us whether the stored value is a zero or one, but for the MLC bucket, the answer to that test is not enough. Based on the first result, we need to perform a second test to see if the bucket is above or below 25%/75% full. All this additional testing takes time, which is why SLC reads are faster than MLC reads, which in turn are faster than TLC reads.

4.2   Emerging Technology: Overview of HEMT

The High Electron Mobility Transistor (HEMT) is a form of Field Effect Transistor (FET) that is used to provide very high levels of performance at microwave frequencies.

The HEMT offers a combination of low noise figure combined with the ability to operate at very high microwave frequencies. Accordingly the device is used in areas of RF design, where high performance is required at very high RF frequencies.

HEMT Development

The development of HEMT took many years. The specific mode of carrier transport used in HEMTs was first investigated in the year 1969, but it was not until 1980, that the first experimental devices were available for the latest RF design projects. The usage of HEMT started in 1980s, but in view of their initial high cost their use was considerably limited.

Now with their cost somewhat less, they are widely used, they are now used in mobile telecommunications as well as a variety of microwave radio communication links and many other RF design applications.

HEMT Construction

The key element within a HEMT is the specialized PN junction that it uses. It is known as a heterojunction and consists of a junction that uses different materials on either side of the junction. The most common materials used are Gallium Arsenide (GaAs) and Aluminium Gallium Arsenide (AlGaAs). Generally, GaAs is used, because it provides a high level of basic electron mobility and this is crucial to the operation of the device. Silicon has a much lower level of electron mobility and as a result it is never used in a HEMT.

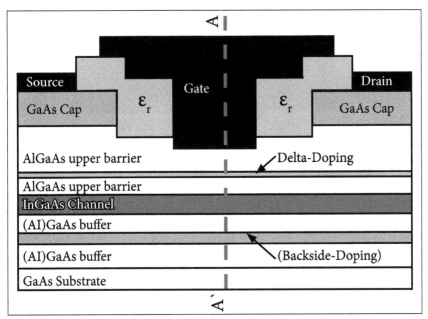

Schematic cross-section of HEMT.

There are a variety of different structures that can be used within a HEMT, but basically all use the same manufacturing process.

In the manufacture of a HEMT, first an intrinsic layer of GaAS is set down on the semi-insulating GaAS layer. This is only about one micron thick. Next, a very thin layer, between 30-60 Angstroms, of intrinsic AlGaAs is set down on top of this. Its purpose is to ensure the separation of the heterojunction interface from the doped AlGaAs region. This is critical, if high electron mobility is to be achieved.

The doped layer of AlGaAs, about 500 Angstroms thick, is set down above this as shown in the above figure. Precise control of thickness of this layer is required and special techniques are required for the control.

There are two main structures that are used. These are:

- The self-aligned ion implanted structure

- The recess gate structure

In case of self-aligned ion implanted structure, the gate, drain and source are set down and are generally metallic contacts, although source and drain contacts may sometimes be made from germanium. Generally, the gate is made from titanium and it forms a minute reverse biased junction similar to that of the GaAs FET.

For the recess gate structure, another layer of n-type gallium arsenide is set down to enable the drain and source contacts. Areas are etched as shown in the figure. Also, the thickness under the gate is very critical, since the threshold voltage of FET is determined by this. Hence, the size of gate and channel is very small. Typically, the gate is only 0.25 microns or less, enabling the device to have a very good high frequency performance.

HEMT Operation

The operation of the HEMT is somewhat different from other types of FET and as a result, it is able to give a very much improved performance over the standard junction or MOSFETs and in particular, in microwave radio applications.

Electrons from n-type region move through the crystal lattice and many remain close to the heterojunction. These electrons form a layer that is only one layer thick, forming what is known as a two dimensional electron gas. Within this region, the electrons are able to move freely, because there are no other donor electrons or other items with which the electrons will collide and the mobility of electrons in the gas is very high.

A bias applied to the gate formed as a Schottky barrier diode is used to modulate the number of electrons in the channel formed from the 2D electron gas and in turn this controls the conductivity of the device. This can be compared to the more traditional types of FET, where the width of the channel is changed by the gate bias.

Applications

The HEMT was originally developed for high speed applications. When the device was first fabricated, it exhibited a very low noise figure. This is related to the nature of the two dimensional electron gas and the fact that there are less electron collisions.

As a result of their noise performance, they are used in low noise small signal amplifiers, power amplifiers, oscillators and mixers operating at frequencies up to 60 GHz and more and it is anticipated that, ultimately devices will be widely available for frequencies up to about 100 GHz. In fact, HEMT devices are used in a wide range of RF design applications including cellular telecommunications, Direct Broadcast Receivers (DBS), radar, radio astronomy and any RF design applications that require a combination of low noise and very high frequency performance.

HEMTs are manufactured by many semiconductor device manufacturers across the globe. They may be in the form of discrete transistors, but nowadays they are more incorporated into the integrated circuit. These Monolithic Microwave Integrated Circuit chips (MMICs) are widely used for RF design applications. HEMT based MMICs are widely used to provide the required level of performance in many areas.

4.2.1 FinFET

FINFET technology takes its name from the fact that the FET structure looks like a set of fins when viewed. The main characteristic of the FINFET is that, it has a conducting channel wrapped by a thin silicon fin, from which it gains its name. The thickness of the fin determines the effective channel length of the device.

In terms of its structure, it typically has a vertical fin on the substrate, which runs between a larger drain and source area. This protrudes vertically as a fin above the substrate.

The gate orientation is at right angles to the vertical fin and to traverse from one side of the fin to the other, it wraps over the fin, enabling it to interface with three sides of the fin or channel. This form of gate structure provides an improved electrical control over the channel conduction and it helps to reduce the leakage current levels and overcomes some other short channel effects.

The term FINFET is sometimes used to describe any of the fin based, multigate transistor architecture regardless of the number of gates.

By modifying the MOSFET architecture and using multiple gates, short channel effects and off-state leakage current can be suppressed much more effectively than by forming an SOI (Silicon On Insulator) substrate multigate transistors.

The trigate or FINFET is shown in the figure below. It allows for further channel length scaling and it may even supersede the SOIMOSFET.

The distinguishing characteristics of the FINFET is that, the gate straddles a thin, fin shaped body forming three self-aligned channels along the top and vertical sidewall surfaces of the fin. The use of three gates around the fin ensures excellent electrostatic control.

As the channel length is scaled down, short channel effects and off-state leakage current are suppressed by scaling down the width of the fin. The fin width should be roughly half the channel length i.e., for a 22 mm node, a fin width of 15 mm is required.

Thus, the width of the fin which is defined by the lithography process is the minimum dimension in a FINFET. The vertical nature of a FINFET provides a greater device width per wafer area enabling the FINFET to be packed more densely than the planar MOSFETs.

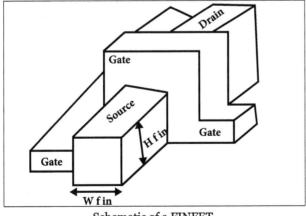

Schematic of a FINFET.

However, the fabrication of a uniform narrow fin is one of the primary challenges in fabricating FINFETs. Regarding the device performance, narrow fins create a high access resistance, reducing the ON current. Another challenge is the implementation of high levels of strain into the fins to boost the ON current.

4.2.2 Organic FET (OFET)

Our daily life involves the continuous use of electronic devices like TV, ATM cards, computer screens, etc. Since the invention of the first transistor in 1947, the majority of these devices have mainly been based on inorganic semiconductors and in particular, on silicon. However, due to technological limitations associated with the use of silicon, substantial efforts are currently devoted to develop organic electronics.

The processing characteristics of organic semiconductors make them potentially useful for electronic applications, where low cost, large area coverage and structural flexibility are required. Organic Field Effect Transistors (OFETs) also known as organic thin film transistors are attracting widespread interest in recent times, because of the possibility to fabricate these with acceptable performances over large areas and on flexible substrates using cost effective and materials efficient fabrication methods. These include the well-known spin casting technique and also newer methods of ink jet or other forms of printing.

Its rapid growth has been spurred primarily by the remarkable development of new materials with improved characteristics and also advancements in understanding their structure-morphology property relations.

Field effect transistors are the main logic units in electronic circuits, where they usually function either as an amplifier or a switch. Organic Field Effect Transistors are mainly based on two types of semiconductor conjugated polymers and small conjugated molecules. The first OFET was reported in 1986 and was based on a film of electrochemically grown polythiophene.

Four years later, the first OFET employing a small conjugated molecule was fabricated. The performance of OFETs in the last 20 years has improved enormously. Nowadays, charge carrier motilities of the same order as amorphous silicon are achieved in the best OFETs.

Thiophene and especially, polyacene derivatives are considered to be the benchmark in OFETs and most of the best motilities have been found in these two family of compounds. However, devices prepared with these molecules are typically prepared by the evaporation of organic materials due to their low solubility in common organic solvents.

Basic Device Structure

Field effect transistors are three terminal device comprising of a gate, source and drain electrodes. In an OFET, an organic semiconductor is deposited to bridge the source and drain electrodes and is itself spaced from the gate electrode by an insulating gate dielectric layer. The organic semiconductor can be a pi-conjugated polymer or oligomer. Important examples include oligothiophenes and polythiophenes and substituted pentacenes, which can be deposited by solution processing.

This solution capability is central to the objective to fabricate large arrays of transistors on large format and potentially flexible substrates, for example, as required in electronic paper and posters.

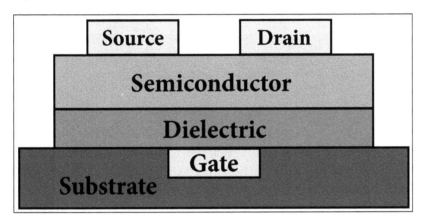

Two voltages are applied relative to the source electrode, which is kept at common (0 V): the drain voltage (V_{ds}) is applied to the drain electrode, whereas the gate voltage (V_{gs}) is applied to the gate electrode. This gate voltage provides an electric field that leads to the accumulation of charge carriers at the semiconductor/dielectric interface which modulates the source to drain conductance. Unlike traditional silicon and other doped inorganic semiconductors, which can operate in both the inversion and accumulation modes, OFETs at present operate in the accumulation mode due to the difficulty to impose a stable controlled level of background doping.

Depending on whether p-channel or n-channel FET characteristics are being measured, the V_{ds} and V_{gs} are swept in the negative or positive voltages to accumulate the appropriate sign of charge carriers at the semiconductor/dielectric interface, i.e., holes or electrons. Then, the measured field effect characteristics of these devices can be classified as transfer characteristics and output characteristics depending on whether V_{gs} is varied while V_{ds} is kept constant or V_{ds} is varied while V_{gs} is kept constant.

Typically, these characteristics are measured on a semiconductor parameter analyzer and sometimes as a function of temperature to understand the underlying physics. For research, devices are fabricated on 200 nm thick thermal oxide on doped silicon substrate as the gate electrode, V_{gs} can be applied routinely up to 60 V without gate dielectric breakdown.

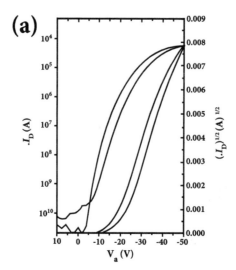

These voltages can be downscaled to a few volts by using thinner insulators and by operating at lower currents. In the transfer curves (a), as the magnitude of V_{gs} increases for a given V_{ds}, the source-drain current increases quadratically beyond a threshold voltage with V_{gs} and then quasi-linearly with V_{gs} as V_{gs} becomes larger than V_{ds}. In this $V_{gs} > V_{ds}$ regime, the linear regime field effect mobility (mFET) can be evaluated using a standard equation from silicon metal oxide semiconductor FET theory.

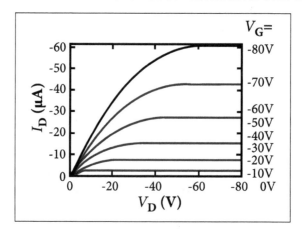

In the output curves (b), as the magnitude of V_{ds} increases for a given V_{gs}, the source-drain current increases linearly with V_{ds} and then levels off i.e., saturates, as V_{ds} becomes larger than V_{gs}. In this $V_{ds} > V_{gs}$ regime, the saturation mFET can similarly be evaluated.

These two mFET values can somewhat differ depending on the quality of the device characteristics. Further, it is appreciated that, these values can also depend strongly on the dielectric interface and the organic semiconductor morphology at this interface. Further they can vary with V_{gs} and temperature and can be degraded by contact resistance. These mFET values are still useful indicators of performance of the organic semiconductor in particular device configurations.

There are two common device configurations used in OFETs. They are:

- Top contact

- Bottom contact

Top Contact

In top contact, the source and drain electrodes are evaporated on the top of the organic material.

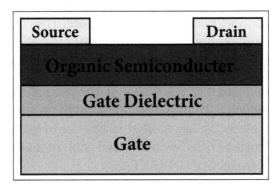

Bottom Contact

In bottom contact, the source and drain electrodes are evaporated on the dielectric before depositing the organic semiconductor.

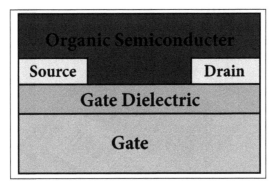

4.2.3 Graphene Nano-Ribbon Field Effect Transistor (GNRFET)

The Graphene Nano-Ribbon Field Effect Transistor (GNRFET) is an emerging technology that received much attention in recent years. Recent work on GNRFET circuit simulations has shown that GNRFETs may have potential in low power applications.

Graphene, a single atomic layer of graphite, is a huge material that experiences a new path in electronic applications. Due to its ultra-thin body and high carrier mobility, graphene has attracted significant attention as a material for next generation technologies that enhance the future high speed nano-electronic devices.

Graphene is a tremendous substance as a channel material for future high speed field effect transistor performance beyond dimensional scaling. The band-gap of this material is zero. When patterned into Nano-scale ribbons, band-gap opens due to lateral quantum confinement.

The narrow stripes of graphene with width less than 100 nm, known as Graphene Nano-Ribbons (GNRs) have finite energy gaps. The GNRs display an impressive current carrying capacity of more than 108 A/cm² for the widths less than 16 nm. Further, the breakdown voltage is estimated to be around 2.5 V for GNRs with widths of 22 nm.

The GNRs are used as channel materials on transistors. Transistors made of GNRs are called Graphene Nano-Ribbon Field Effect Transistors (GNRFETs), which are the potential alternatives to CMOS devices both experimentally and theoretically.

One of the critical issues related to transistors is the ratio of ON state current to OFF state current $\left(I_{on} / I_{off}\right)$ which is an important parameter on switching performances of the device. Much of the attention has shifted to the use of graphene in RF transistors with large cut-off frequencies $\left(f_{T}\right)$. Temperature is one of the device parameter with a major effect on the performance of GNRFET.

4.3   IC Design for Internet of Everything (IoE)

Analog Integrated Circuits

Analog IC.

Analog integrated circuits were primarily designed using hand calculations before the invention of microprocessors and other software dependent design tools. Analog IC design is used for designing operational amplifiers, active filters, linear regulators, oscillators and phase locked loops. The semiconductor parameters such as gain, power dissipation and resistance are more concerned in the designing of analog IC.

Analog IC Design

Analog IC design process includes system design, circuit design, component design, circuit simulations, system simulations, integrated circuit layout design, interconnect, verification, fabrication, device debug, circuit debug and system debug. Digital IC design can be automated, but analog IC design is very difficult, challenging and cannot be automated.

The practical analog IC design involves the following steps:

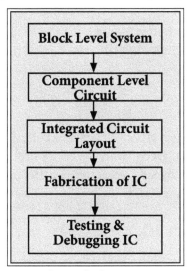

Analog IC design process.

Block Level System

Primarily the ideas are implemented for designing the block level design for the desired analog integrated circuit. Different blocks are designed and connected to obtain a complete block level system.

Component Level Circuit

Based on the block level system, different suitable components are used and connected so as to form a component level circuit. Using this circuit as the basic circuit for analog IC design, it is used for simulation.

The component level circuit is used for verification. This circuit design is simulated based on the simulation results and the component level circuit of the analog integrated circuit is verified.

Integrated Circuit Layout

After the verification of component level circuit of analog integrated circuit using simulations, analog IC layout is designed using the physical translation. Thus, an analog IC layout is designed.

Fabrication of IC

Fabrication of analog IC involves several steps such as creating semiconductor wafer using semiconductor material. Integrating different electrical and electronics components such as resistors, transistors, etc., on the wafer and packing the chip to form a package IC are done.

Testing and Debugging IC

The analog IC is then tested and debugged with the estimated results. Then IC prototype is designed and used for characterizing the IC and the evaluation board is used for evaluating the analog IC.

Applications of Analog IC

There are different examples for analog integrated circuit designs such as operational amplifiers, power management circuits and sensors that are used with continuous signals for performing the functions such as active filtering, mixing, power distributing for components within the chip and so on.

4.3.1 Digital and Memory IC

Digital ICs operate at only a few defined levels or states, rather than over a continuous range of signal amplitudes. These devices are used in computers, computer networks, modems and frequency counters. The fundamental building blocks of digital ICs are logic gates, which work with binary data i.e., signals that have only two different states called low (logic 0) and high (logic 1).

Digital circuits are circuits dealing with signals restricted to the extreme limits of zero and some full amount. This stands in contrast to analog circuits, in which signals are free to vary continuously between the limits imposed by power supply voltage and circuit resistances. These circuits find use in true/false logical operations and digital computations.

Here we make use of the integrated circuit components. Such components are actually networks of interconnected components manufactured on a single wafer of semiconducting material. Integrated circuits providing a multitude of pre-engineered functions are available at very low cost, benefitting students, hobbyists and professional circuit designers. Most integrated circuits provide the same functionality as discrete semiconductor circuits at higher levels of reliability and cost.

Here we will primarily use CMOS technology, as this form of IC design allows for a broad range of power supply voltage, while maintaining generally low power consumption levels. Though CMOS circuitry is susceptible to damage from static electricity, modern CMOS ICs are far more tolerant than CMOS ICs of the past, in terms of electrostatic discharge, reducing the risk of chip failure by mishandling.

Proper handling of CMOS involves the use of anti-static foam for storage and transport of IC's and measures to prevent static charge from building up on our body. Circuits using TTL technology require a regulated power supply voltage of 5 V and will not tolerate any substantial deviation from this voltage level.

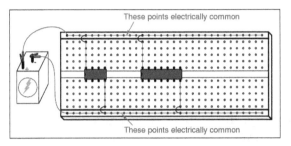

When building digital circuits using IC chips, it is highly recommended to use a breadboard with power supply rail connections along the length. These are sets of holes in the breadboard that are electrically common along the entire length of the board. Connect one to the positive terminal of a battery and the other to the negative terminal and DC power will be available to any area of the breadboard through short jumper wires.

With many of these ICs having "enable", "disable" and "reset" terminals needing to be maintained in a "high" or "low" state, not to mention the V_{DD} or V_{CC} and ground power terminals which require connection to the power supply, having both the terminals of the power supply readily available for connection at any point along the board's length is very useful.

Most of the breadboards have these power supply rail holes, but some do not. However, digital circuits seem to require more connections to the power supply than other types of breadboard circuits, making this feature more convenient.

Memory IC

Nowadays all the semiconductor memory devices are now available in Integrated Circuit (IC) form. Each memory IC can store a large number of words. Memory ICs are available in various sizes. The examples of ICs are 64 × 4, 256 × 8, 1 K × 8, 1 M × 8, etc.

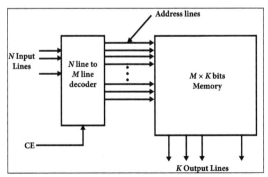

(1) Block diagram of M × K bits memory.

Each memory IC should have address and data lines including chip select $\overline{CS}$, output enable $\overline{OE}$ and Read/Write $R/\overline{W}$ control signals. Figure (2) shows the memory organization of an IC. This chip has address lines, data lines, chip select signals and read and write control signals.

Address Lines

The memory ICs should have address lines to receive the address values. For a 1K byte memory, ten address lines $A_0 - A_9$ exist. The relationship between the number of address lines and size of memory is 2^n, where n is the number of address lines. Similarly, for a 64K byte memory, the number of address line is 16, A_0 to A_{15}. Address line $A_0 - A_{n-1}$ can be used to select one of the 2n memory locations.

Data Lines

Data lines provide data input to the IC during write operation and data will be output from IC during read operation. M data lines $D_0 - D_{m-1}$ are used to transfer data between microprocessor and memory IC.

Chip Select Signal $\overline{CS}$

The chip select signal $\overline{CS}$ can enable the chip. When $\overline{CS}$ is low, memory access within the chip is possible.

Read or Write $R/\overline{W}$

The read or write operation can be performed based on $R/\overline{W}$ control signal. If $R/\overline{W} = 1$, data will he read from the memory. When $R/\overline{W} = 0$, data will be stored in the memory IC.

Outputs Enable $\overline{OE}$

The output enable signal is used to connect the output with the data bus.

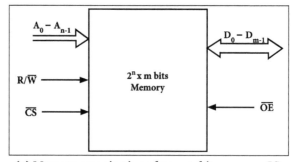

(2) Memory organization of 2n × m bits memory IC.

For example, the memory organization of 256 × 4 memories IC is depicted in figure (3). This memory IC has 8 address lines $A_0 - A_7$, to select 256 memory locations. As there are four data lines, 4 data bits can be stored in each location. Therefore, the size of the memory is 256 × 4 bits.

Memory ICs are available in four bit and eight bit word configurations. In some applications, 16 bits and more than 16 bits are also used. The memory capacity of each IC is limited. Therefore, memory expansion is required. The memory size can be expanded by increasing the word size and address locations. The memory expansion can also be possible by proper interconnections of decoder and memory ICs.

Figure (4) shows 2K × 8 bits memory using two 1K × 8 bits. A 2K byte RAM can be developed using two 1K byte RAM ICs. In this case, $\overline{CS}$ is directly connected with IC1 and the complement of $\overline{CS}$ is connected to IC2. R/$\overline{W}$ and $\overline{OE}$ control signals of both ICs are directly interconnected as shown in figure (4). The address lines $A_0 - A_9$ of the ICs are connected in parallel. Chip-1 provides the 1K addresses from 0 to 1023 and chip-2 provides the next 1K addresses from 1024 to 2047.

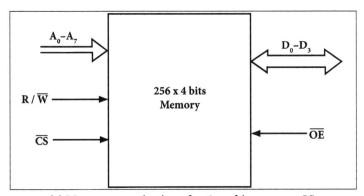

(3) Memory organization of 256 × 4 bits memory IC.

As for the first 1K addresses, chip-1 is activated and for the next 1K addresses, chip-2 is activated. The chip select signal is connected with the address line A_{10}. Each chip provides 8 data lines $D_0 - D_7$. So that memory size is increased from 1K byte to 2K byte.

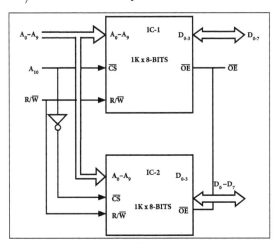

(4) Memory organization of 2K × 8 bits memory using two 1K × 8 bits.

Another example is that two 1K × 4 bits RAMs can be combined to develop 1K bytes RAM as shown in figure (5). IC-1 and IC-2 have ten address lines, which are connected in parallel. The chip select $\overline{CS}$, read/write R/$\overline{W}$ and output enable $\overline{OE}$ are also connected together. In this case, memory size is fixed, but word size is increased from 4 bit to 8 bit. IC-1 and IC-2 are selected at a time for 8-bit data storage or data read operation.

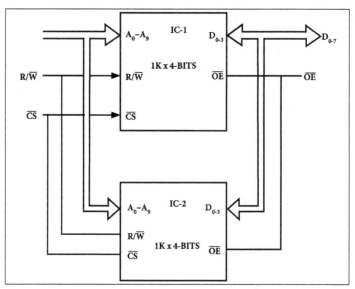

(5) Memory organization of 1K × 8 bits memory using two 1K × 4 bits.

4.3.2 Mixed-Signal IC

Mixed-signal ICs are chips that contain both digital and analog circuits on the same chip. This category of chip has grown dramatically with the increased use of 3G cell phones and other portable technologies.

They are often used to convert analog signals to digital signals, so that digital devices can process them. For example, mixed-signal ICs are essential components for FM tuners in digital products such as media players, which have digital amplifiers. Any analog signal can be digitized using a very basic Analog to Digital Converter (ADC) and the smallest and most energy efficient of these would be in the form of mixed-signal ICs.

These ICs also enable technologies like power over Ethernet. The analog power signal, in this case, a 60 Hz, 120 V AC current, is transmitted along a digital data signal over the same wire. Mixed-signal ICs allow this.

They are more difficult to design and manufacture. For example, an efficient mixed-signal IC would have its analog and digital components share a common power supply. However, analog and digital components have different power needs and consumption characteristics, which make this a non-trivial goal in chip design.

Mixed-Signal Processing Block

A typical mixed-signal processing block consists of both analog and digital blocks. The first step in the design of a mixed-signal system is to decide the portioning of the analog and digital sections. A typical block diagram of signal processing system is depicted in the below figure.

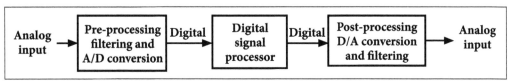

A typical block diagram of signal processing system.

Generally, the input signal is analog. The first block is the pre-processing block and consist of filter, Automatic Gain Control (AGC) circuit and an Analog to Digital Converter (ADC), very strict speed and accuracy requirements are placed on the components in this block.

The digital output of the first block is given as input to the second block, which is called as a digital signal processor. The advantage of digital signal processing is the ease of implementation of circuits. The final output required is analog. The third block consists of Digital to Analog Converter (DAC), amplifier and filter.

Example of mixed-signal:

The below figure shows the block diagram of a digital read/write channel for disc recording application.

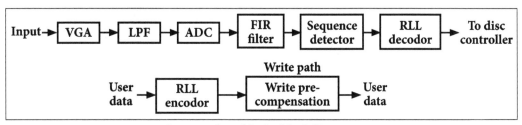

Integrated read/write block diagram.

This IC receives an analog signal from an external pre-amplifier, which senses magnetic transitions on a disc. The read signal is first amplified and then passed through a low pass filter. This analog signal is given to an ADC. The digitized output of the ADC is filtered by an FIR filter.

The sequence detector anticipates linear inter-symbol interference and after processing the received sequence of values deduces the most likely data. The bit stream from the sequence detector is decoded by the Run Length Limited (RLL) decode block and the bit stream appears on the read channel output pins. In write mode, data is first encoded by a RLL encoder block. The encoded data is then passed to the write pre-compensation circuitry.

Uses of Mixed-Signal

- Clocking and timing circuits

- Digital input/output

- Supply voltage regulation

- Sensor interfacing

- Wireless communication

4.3.3 RF/MM-Wave

RF, Micro and Millimeter waves constitute a vital area of electrical engineering encompassing design, modeling, simulation, experimentation and analysis of single devices, circuits and packaging with applications to communications, imaging, radar systems and basic science.

Terahertz IC

The Terahertz Monolithic Integrated Circuit (TMIC) bridges the gap using a super-scaled, 25 nm gate length, indium phosphide, high electron mobility transistor. The transistor measures a gain of 10 dB at 1 THz and 9 dB at 1.03 THz.

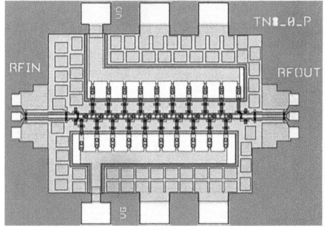

TMIC.

Terahertz Technology

Terahertz (THz) technology has gained high interests in various applications owing to its unique characteristics. Generally, the terahertz range is widely conceived as 0.3 THz to 3 THz, whose wavelength (λ) is between 1-0.1 mm, a transition region between electronics and photonics. It has great potential in sensing and communication applications. Terahertz radiation can penetrate dielectric materials without causing any destructive ionization of the material.

In sensing applications, the terahertz imager can achieve much higher imaging resolution than millimeter wave counterpart. Terahertz radiation has been widely used in spectroscopy by using the vibration of molecules for a given terahertz radiation frequency, which can produce a unique fingerprint depending on type of dielectric materials. Biomedical spectroscopy and remote gas sensing are the promising examples of the application.

Owing to extremely short wavelength, the terahertz compact range is useful for indoor millimeter wave Radar Cross-Section (RCS) characterization of tanks and aircrafts. The terahertz spectrum has great potential in ultrafast wireless communication by providing wide bandwidth in the new spectrum regime.

In order to expand the use of electromagnetic spectrum, designing highly efficient compact sources and detectors are essential. However, there exist many challenges in achieving compact, reliable sources and detectors in this transition spectrum regime.

Until recently, fully integrated THz transceiver in silicon has not been considered as a promising solution, due to limited device performance and large propagation losses coupled through a resistive silicon substrate in terahertz range. However, we have witnessed revolutionary achievements in RF and millimeter wave integrated circuit technology during the last decade with advancement of the nano-scale silicon technology.

Considering the impact of RFICs in its compactness, low cost and mass production, the THz IC will open a new era in imaging, sensing, spectroscopy and ultrafast wireless communication.

Terahertz Regime

The terahertz range usually implies the unique frequency range, which lies between the microwaves and infrared in the electromagnetic spectrum as shown in the below figure. Owing to this loosely defined range, slightly different frequency ranges are considered as terahertz regime. Roughly, it ranges between 0.1-30 THz. Sometimes, it ranges between 0.3-10 THz, while microwave electronics considers the terahertz range as 0.3-3 THz, which is termed as sub-millimeter wave range.

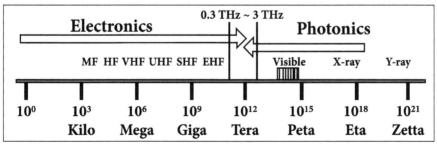

Unique terahertz region placed in a transition region between microwaves and infrared in the electromagnetic spectrum.

Applications

Applications include atmospheric sensing, high resolution security imaging systems, radio astronomy, medical imaging and improved collision avoidance radar and communications network with greater capacity. It could also help improve system range and reduce size, weight and power consumption of the existing systems.

Permissions

Index

Printed in the USA
CPSIA information can be obtained
at www.ICGtesting.com
JSHW051352091023
49903JS00006B/120

9 781647 254292